WORLD HEALTH ORGANIZATION

INTERNATIONAL AGENCY FOR RESEARCH ON CANCER

LABORATORY DECONTAMI-NATION AND DESTRUCTION OF CARCINOGENS IN LABORATORY WASTES: SOME *N*-NITROSAMINES

EDITORS

M. CASTEGNARO, G. EISENBRAND, G. ELLEN,
L. KEEFER, D. KLEIN, E.B. SANSONE, D. SPINCER,
G. TELLING & K. WEBB

TECHNICAL EDITOR FOR IARC

W. DAVIS

IARC SCIENTIFIC PUBLICATIONS No. 43

INTERNATIONAL AGENCY FOR RESEARCH ON CANCER
LYON
1982

The International Agency for Research on Cancer (IARC) was established in 1965 by the World Health Assembly, as an independently financed organization within the framework of the World Health Organization. The headquarters of the Agency are at Lyon, France, and it has Research Centres in Iran, Kenya and Singapore.

The Agency conducts a programme of research concentrating particularly on the epidemiology of cancer and the study of potential carcinogens in the human environment. Its field studies are supplemented by biological and chemical research carried out in the Agency's laboratories in Lyon and, through collaborative research agreements, in national research institutions in many countries. The Agency also conducts a programme for the education and training of personnel for cancer research.

The publications of the Agency are intended to contribute to the dissemination of authoritative information on different aspects of cancer research

ISBN 92 8 321143 X

© International Agency for Research on Cancer 1982

150 cours Albert-Thomas, 69372 Lyon Cédex 08, France

PRINTED IN SWITZERLAND

CONTENTS

Foreword v

Preamble vii

N-Nitrosamines studied 1

Introduction

    Possible human health hazards from *N*-nitrosamines 3
    Analysis: (a) NDMA, NDEA, NDPA, NDBA, NPIP, NPYR and NMOR   4
            (b) *N*,*N'*-Dinitrosopiperazine 4

Degradation techniques

    Methods investigated 5
    Some additional methods for investigating the decontamination
      of laboratory wastes 5

Recommended methods of degradation

    Introduction 9
    Collaborating organizations 9
    Methods index 13
        Method 1. Destruction of *N*-nitrosamines in laboratory
               wastes using denitrosation with hydrobromic
               acid 15

        Method 2. Destruction of *N*-nitrosamines in laboratory
               wastes using potassium permanganate 23

        Method 3. Decontamination of *N*-nitrosamine contaminated
               glassware 27

        Method 4. Decontamination of laboratory wastes using
               triethyloxonium salts 31

        Method 5. Destruction of *N*-nitrosamines in laboratory
               wastes using nickel-aluminium alloy in
               potassium hydroxide solution 35

Appendix A - Nomenclature and chemical and physical data . . . 43

    1.1 N-Nitrosodimethylamine 43
    1.2 N-Nitrosodiethylamine 44
    1.3 N-Nitrosodipropylamine 45
    1.4 N-Nitrosodibutylamine 46
    1.5 N-Nitrosopiperidine 47
    1.6 N-Nitrosopyrrolidine 48
    1.7 N-Nitrosomorpholine 49
    1.8 N,N'-Dinitrosopiperazine 50

    2.   Spectral properties 51

Appendix B - Further studies relevant to degradation of
            N-nitrosamines

    1.   Biological degradation 53
    2.   Chemical reactions 54

References 63

FOREWORD

Chemical and biological research, involving the formation and reactions of N-nitrosamines, occupies a major place in the programmes of laboratories in many different countries. The compounds and their precursors are widespread in the human environment, and although as yet no human cancer has been causally associated with N-nitrosamine exposure, the importance of studies of their carcinogenicity cannot be overrated.

The International Agency for Research on Cancer has carried out research itself in this field for several years now, and has been acutely aware of the problems of safety involved in the handling of N-nitroso compounds. We were, therefore, very pleased to have the support of the Division of Safety, National Institutes of Health (USA) to develop a collaborative programme on methods of destruction of carcinogenic laboratory wastes, including those derived from the N-nitrosamines.

The publications of the series of such methods will, it is hoped, contribute to greater safety in and around laboratories engaged in this necessary and important field of research.

L. TOMATIS, M.D.

Director
International Agency
for Research on Cancer
Lyon, France

PREAMBLE

Biomedical research involving chemical carcinogens inevitably results in the production of waste products containing chemical carcinogens. These waste products range from those minimally contaminated with carcinogens such as spent fluids from cell culture applications and carcasses of animals used in bioassay projects to unused pure compounds no longer required after the completion of a research program. During the research process, equipment, work surfaces, and laboratory spaces may also become contaminated with chemical carcinogens through unavoidable contact or accidental exposure. The production of these waste products and potential for contamination require that effective methods for the destruction of chemical carcinogens, designed specifically for laboratory application, be developed and used to ensure safe and environmentally sound disposal practices.

The International Agency for Research on Cancer, with the support of the Division of Safety, National Institutes of Health, has undertaken a special program to develop, validate and publish methods for the destruction and disposal of laboratory waste containing carcinogens. This monograph on N-nitrosamines is the second volume of a series that will eventually include monographs on N-nitrosamides, polycyclic aromatic hydrocarbons, aryl halides, halogenated ethers, aromatic amines and hydrazines.

An important feature of this special program is to establish a base of information from which appropriate methods can be developed and validated. The International Agency for Research on Cancer and the Division of Safety encourage individual scientists and laboratories to participate in this endeavor by contributing suggested methods and participating in collaborative validation studies. It is our hope that this special program will serve as a catalyst for stimulating research in this area and for sharing the results of such investigations. We all share a common responsibility for the safe disposal of laboratory waste. Let us pool our resources and talents in a common effort to help ensure that effective methods are available for the destruction and disposal of laboratory waste containing chemical carcinogens.

W. Emmett Barkley, Ph.D.
Director, Division of Safety,
National Institutes of Health

N-NITROSAMINES STUDIED

The following *N*-nitrosamines have been considered for this document. It is anticipated that methods for their destruction will be applicable to *N*-nitroso compounds in general.

N-Nitrosamine	Chemical Abstracts Services Registry No.	Chemical Abstracts name	Abbreviation used
N-Nitrosodimethylamine	62-75-9	*N*-Methyl-*N*-nitrosomethanamine	NDMA
N-Nitrosodiethylamine	55-18-5	*N*-Ethyl-*N*-nitrosoethanamine	NDEA
N-Nitrosodipropylamine	621-64-7	*N*-Nitroso-*N*-propyl-1-propanamine	NDPA
N-Nitrosodibutylamine	924-16-3	*N*-Butyl-*N*-nitroso-1-butanamine	NDBA
N-Nitrosopiperidine	100-75-4	1-Nitroso-piperidine	NPIP
N-Nitrosopyrrolidine	930-55-2	1-Nitrosopyrrolidine	NPYR
N-Nitrosomorpholine	59-89-2	4-Nitrosomorpholine	NMOR
N,*N*'-Dinitrosopiperazine	140-79-4	1,4-Dinitrosopiperazine	Di-NPZ

INTRODUCTION

Possible human health hazards from N-*nitrosamines*

N-Nitrosamines are products formed from the nitrosation of secondary amines, but are occasionally also derived from the nitrosation of some primary and tertiary amines or quaternary ammonium compounds.

Man may be exposed to N-nitrosamines from various sources: foodstuffs, particularly those processed with nitrate and nitrite, and a variety of alcoholic drinks may lead to an intake not likely to exceed 1 µg/day (Gough et al., 1978; Spiegelhalder et al., 1980; Stephany & Schuller, 1980). Other potential sources of exposure include the use of cosmetics, certain types of drugs, pesticides or other industrial chemicals and chewing and smoking of tobacco. It has been shown that N-nitrosamines frequently occur (sometimes in ppm levels) at the work place in rubber, tanning and metal-working industries (Fajen et al., 1979, 1982; Spiegelhalder & Preussmann, 1982). Additionally, N-nitrosamines are probably also formed in man through the intake of nitrosatable precursors and nitrate. Considering these possible sources of exposure, it is therefore important that work in the laboratory should not further add to the total carcinogenic burden.

The carcinogenicity of N-nitroso compounds has been extensively documented in animals (Magee et al., 1976) and up to now tumours have been produced in thirty-nine animal species including monkeys, birds, fish and amphibia (Bogovski & Bogovski, 1981). In strain A mice lifetime exposure to ppb levels of N-nitrosodimethylamine in the drinking water caused an increase in the incidence of lung cancer (pulmonary adenoma), showing the effectiveness of this group of compounds as carcinogens. The sites at which tumours appear varies in a remarkable fashion with the chemical structure, and they have been induced in virtually all tissues and organs. In rats, for example, the organs most frequently effected by N-nitrosamines are, in descending order, liver, oesophagus, nasal cavity, kidney (Bartsch et al., 1971).

N-Nitrosamines are also mutagenic in a variety of bacterial and mammalian test systems *in vitro* and *in vivo*, and some of them are also teratogenic (Montesano & Bartsch, 1976).

Although there is no direct evidence for the carcinogenicity of
N-nitrosamines in human beings, it is known that the pathological
changes in the liver, after poisoning of human subjects by N-nitroso-
dimethylamine, are the same as in experimental animals, the main lesion
being centrilobular necrosis. Furthermore, the metabolic pathways of
N-nitrosamines leading to reactive intermediates and the resulting DNA-
binding products have been found to be similar in human and animal
livers.

For the following N-nitrosamines: N-nitrosodi-n-butylamine,
N-nitrosodiethanolamine, N-nitrosodiethylamine, N-nitrosodi-n-propyl-
amine, N-nitrosomethylethylamine, N-nitrosomethylvinylamine, N-nitroso-
morpholine, N-nitrosopiperidine, N-nitrosopyrrolidine and N-nitroso-
nornicotine there is sufficient evidence of carcinogenicity in experimental animals (IARC, 1979a). For these reasons, N-nitrosamines can be
regarded for all practical purposes as though they pose a carcinogenic
risk to humans.

Because at present there is insufficient knowledge to make reliable
estimates of acceptable levels (thresholds) of carcinogens, including
N-nitrosamines, which are not effective in inducing tumours, minimizing
exposure of human subjects is therefore a particularly necessary measure.
It is hoped that the methods for the destruction of N-nitrosamines
described in this volume, will help to avoid health hazards posed by
this class of carcinogens.

Analytical methods

   (a) NDMA, NDEA, NDPA, NDBA, NPIP, NPYR and NMOR

A number of methods have been elaborated for the analysis of these
N-nitrosamines in various substrates (food or drink). They have been
reviewed by Wasserman (1972), Preussmann and Eisenbrand (1972), Crosby
and Sawyer (1976), Walker (1980). Some of these methods, tested in
collaborative studies, proved to be satisfactory (Walker & Castegnaro,
1974, 1976; Castegnaro & Walker, 1978, 1980). As a result of these
studies a Manual, *Environmental Carcinogens. Selected Methods of
Analysis* (Egan et al., eds, 1978) has been published. As the analytical
technology has improved greatly during the last three years, a second
volume of the Manual Series is now being published (Egan et al., eds, 1982).

   (b) N,N'-Dinitrosopiperazine

As no evidence has been found for its presence in the environment,
little effort has been made to elaborate methods of analysis for this
compound.

In the present studies a gaschromatographic method is described,
other methods might be found in the literature (Sander et al., 1973,
1974; Sen & Donaldson, 1974; Rao, 1977; Rao & MacLennon, 1977).

DEGRADATION TECHNIQUES

Methods investigated

Some methods which might be applied to the destruction of N-nitroso compounds have been reviewed by Douglass et al. (1978). They include denitrosation by hydrobromic acid (HBr) and acetic acid, photolytic decomposition by ultraviolet light (UV) and various reduction techniques. Table 1 summarizes methods which have been proposed or evaluated for decontamination of various contaminated wastes.

Some additional methods for investigating the decontamination of laboratory wastes

Of the chemical reactions listed in Appendix B, several appear to merit investigation as decontamination procedures. N-Nitrosamines can be oxidized to nitramines using trifluoroacetic acid/hydrogen peroxide; and reportedly by chromic/sulfuric acids. They may be reduced electrochemically or chemically to amines, hydrazines, nitrosohydroxylamine or ammonia. They can be denitrosated by hydrobromic acid in glacial acetic acid or catalytically at high temperature, e.g. in the Thermal Energy Analyzer (TEA) or the Coulson detector. Destruction of the N-nitroso function can be achieved by treatment with 2,4-bis(4-methoxyphenyl)-1,3,2,4-dithiadiphosphetane-2,4-disulfide (LR) (Jørgensen et al., 1980a, b).

Since it is important that the product of degradation should have a minimum adverse biological effect (IARC, 1979b), and N-nitrodimethylamine being carcinogenic (Druckrey et al., 1967; Goodall & Kennedy, 1976), oxidation of N-nitrosamines to N-nitramines should be rejected as a method of decontamination. Some other methods of possible interest include the formation of a non-volatile oxonium salt as a possible intermediate in a degradation process, destruction of the N-nitroso function by LR or UV treatment. Compounds containing N-N bonds burn two to four times more quickly than compounds with C-N bonds (Fogel'Zang et al., 1974), suggesting that incineration may be an effective way of decontamination.

The preferred route for the listed N-nitrosamines is offered by their cleavage or reduction to amines since these are not reported to be carcinogenic, however, one should remember that this does not apply to N-nitrosamines containing an aryl group as these compounds could produce aromatic amines.

Table 1. Methods of degradation of N-nitrosamines which have been studied

Medium	N-Nitrosamines studied	Degradation methods	Efficiency	Reference
Solutions[a]	NDMA, NDEA, NPIP, NPYR	Reduction with sodium hydroxide (12.5 g/L) and aluminium foil (10 g/L). A system for large volumes of solution is described	On a 15 mg/L solution For NDMA and NDEA > 99.5% For NPIP and NPYR 97%	Gangolli et al. (1974)
Solution[a]	NDMA	Reduction in 1 M potassium hydroxide solution containing 2 M equivalent of aluminium foil leads mainly to formation of the corresponding hydrazine		Emmett et al. (1979)
Solutions[a]	NDMA, NDEA, NDPA	Action of sodium hydroxide (1.3 g/100 ml) and aluminium foil (1 g/100 ml) on an ethanolic solution (30% ethanol) of nitrosamine. The hydrazine can be converted to hydrazone.	NDMA, NDEA, NPIP > 99.7% reduction NDPA, NDBA about 85% reduction	Chien & Thomas (1978)
Solutions	NDMA	Evaluation of various factors which influence photodegradation a) acceleration by 10-fold excess of ascorbic acid b) nitrosamine concentration	Half-life of a 10 mM solution reduced from 120 min to 13 min Rate of degradation decreases with increasing concentration	Emmett et al. (1979)
Solutions	Nitrosomethylaniline (NMA)	Colour produced during degradation retards photolysis	Half-life increases by a factor >> 100 when concentration is increased from 10 mM to 100 mM	Emmett et al. (1979)
Solutions	Nitrosodiisopropylamine in presence of N-methyl-aniline and NDMA in presence of N,N-dimethylaniline	Denitrosation in presence of amine results in transnitrosation		Emmett et al. (1979)
	NDMA	The presence of adsorptive particles (e.g. silica gel)	100 mg of silica gel in a 10 mM solution in benzene reduces the percentage of degradation over 50 min from 48% to 30%	
White solid absorbent matrix	NMA	The depth of contamination below the surface irradiated	For white surface: 80% on surface decreasing rapidly with depth (probably even less efficient for dark matrix)	Emmett et al. (1979)

Table 1. (continued)

Medium	N-Nitrosamines studied	Degradation methods	Efficiency	Reference
Solutions	NDMA	Photolytic degradation in aqueous solution catalysed by acid and by the presence of a nitrous acid scavenger such as urea. Two types of equipment for large volumes are described.	Half-life listed under various conditions	Polo & Chow (1976)
Solutions	NDPA	Photolytic degradation in aqueous solution is not affected by pH in the range 3-9 and produces n-propylamine and di-n-propylamine	77-78% in 2 hr	Saunders & Mosier (1980)
Solutions	NDPA	Treatment with various acids under various conditions. Hydrogen chloride gas, hydrochloric acid and hydrobromic acid are effective in removing and destroying NDPA. Concentrated sulfuric acid acts as a trap but does not destroy NDPA.	Level reduced from 68 ppm to <1 ppm in 20 min by >33% hydrochloric acid or 48% hydrobromic acid and in 2 hr by hydrogen chloride gas at 35 ml/min	Eizember et al. (1979)

[a] There appears to be conflicting evidence concerning the products of reduction from N-nitrosamines (see also in Chemical Reactions: Lund, 1957).

RECOMMENDED METHODS OF DEGRADATION

Introduction

Eight methods proposed in the literature or suggested by workers in this field, have been tested. One, using sodium hypochlorite, has been rejected as it was found to be totally inefficient. A method using a mixture of concentrated chromic and sulfuric acids was tested and its efficiency found questionable (Castegnaro et al., 1982). In view of this, and that regulations in force in several countries prohibit the discharge of chromate into sewage, this method has been withdrawn. A method using cuprous chloride and hydrochloric acid gave unsatisfactory results in the present collaborative study. Methods 1, 2, 3 and 4, as described in this section, have been developed by M. Castegnaro, J. Michelon and E.A. Walker, and Method 5 by E. Sansone, G. Lunn and L. Keefer.

It is recognized that Method 4 is subject to limitations due to the lack of long-term stability of triethyloxonium tetrafluoroborate (TEOF). However, in view of the results obtained in the collaborative study of this method, it has been proposed as a possible route to future development in the field of decontamination procedures. Several other more stable oxonium salts are now available on the market and merit evaluation.

Incineration of *N*-nitrosamine contaminated wastes is a method which is widely used. Unfortunately, it has not proved possible to present a validated method since not only the conditions of incineration may vary widely between different installations, but the technical difficulties of testing the flue gases for the possible presence of carcinogens would be considerable.

Collaborating organizations

Collaborative studies of Methods 1, 2, 3, 4 and 5 have been carried out by the following laboratories:

Institute of Toxicology & Chemotherapy
German Cancer Research Center
Postfach 449
6900 Heidelberg 1, Federal Republic of Germany

National Institute of Public Health
P.O. Box 1
3720 BA Bilthoven, The Netherlands

Department de Nutrition et des Maladies Métaboliques
University of Nancy et Groupe INSERM U-59
40 rue Lionnois
54000 Nancy, France

Department of Health & Human Services
National Institutes of Health
Bethesda, MD 20205, USA

Environmental Control & Research Laboratory
NCI - Frederick Cancer Research Facility
P.O. Box B
Frederick, MD 21701, USA

Laboratory of the Government Chemist
Cornwall House, Stamford Street
London, SE1 9NQ, UK

Imperial Group Limited
Raleigh Road
Bristol BS3 1QX, UK

Unilever Research Laboratory
Colworth House
Sharnbrook, Bedfordshire MK44 1LQ, UK

International Agency for Research on Cancer
150 cours Albert Thomas
69372 Lyon Cédex 08, France

For the samples tested, results are given in section 1 of the appropriate methods.

… DESCRIPTION OF METHODS FOR DECONTAMINATION AND
DESTRUCTION OF *N*-NITROSAMINES IN LABORATORY WASTES

METHODS INDEX

Decontamination of media

Waste category	Recommended N-nitrosamine destruction Method no. (in order of preference)
Undiluted N-nitrosamines	1, 5, 2, 4
Spills of aqueous solutions	2
Spills of CH_2Cl_2 or ethanol solutions	1
Aqueous solutions	5, 2, 1, 4
Solutions in CH_2Cl_2	1, 5, 4
Solutions in ethanol	5, 1
Solutions in olive oil	5
Content of petri dishes	5, 1

Decontamination of glassware

Type of contaminant	Recommended decontamination method			
	Method 3			Method 1
	CH_2Cl_2	MeOH	H_2O	
Undiluted N-nitrosamines				X
Aqueous solutions			X	X
Alcoholic solutions	X	X	X	X
Oily solutions				X
CH_2Cl_2	X			X

METHOD 1
DESTRUCTION OF N-NITROSAMINES IN LABORATORY WASTES USING DENITROSATION WITH HYDROBROMIC ACID

1. SCOPE AND FIELD OF APPLICATION

Procedures are specified for the treatment of the following N-nitrosamines or laboratory wastes contaminated with N-nitrosamines: undiluted N-nitrosamines (7.1), solutions in volatile organic solvents (7.2), solutions in ethanol (7.3), aqueous solutions (7.4), content of petri dishes (7.5), spills (7.6), glassware or solid wastes (7.7). The N-nitrosamines specifically investigated were N-nitrosodimethylamine (NDMA), N-nitrosodiethylamine (NDEA), N-nitrosodipropylamine (NDPA), N-nitrosodibutylamine (NDBA), N-nitrosopiperidine (NPIP), N-nitrosopyrrolidine (NPYR), N-nitrosomorpholine (NMOR) and N,N'-dinitrosopiperazine (Di-NPZ).

The method has been collaboratively studied using solutions of NDMA, NDBA and NPYR in water (100 mg/L, 50 mg/L and 100 mg/L respectively), in ethanol (50 mg/L each) and agar spiked with the same three N-nitrosamines, at the levels of 4 mg, 4.8 mg and 4 mg, respectively.

The method affords better than 99% decontamination for all three substrates tested. For recommended application of the method see Methods Index, p. 13. The method has also been collaboratively tested for the same three N-nitrosamines in olive oil. Due to problems encountered with the technique it was decided not to recommend this method for this substrate.

2. REFERENCES

Eisenbrand, G. & Preussmann, R. (1970) Eine neue Methode zur kolorimetrischen Bestimmung von Nitrosaminen nach Spaltung der Nitrosogruppe mit Bromwasserstoff in Eisessig. *Arzneim.-Forsch., 20*, 1513-1517

Preussmann, R. & Eisenbrand, G. (1972) *Problems and recent results in the analytical determination of N-nitroso compounds. Topics in Chemical Carcinogenesis.* University of Tokyo Press, pp. 323-339

Drescher, G.S. & Franck, C.W. (1978) Estimation of extractable N-nitroso compounds at the part per billion level. *Analyt. Chem., 50*, 2118-2121

Johnson, E.M. & Walters, C.L. (1971) The specificity of the release of nitrite from nitrosamines by hydrobromic acid. *Analyt. Lett.*, *4*, 383-386

Castegnaro, M., Michelon, J. & Walker, E.A. (1982) In: Bartsch, H., O'Neill, I.K., Castegnaro, M. & Okada, M., eds, N-*Nitroso Compounds: Occurrence and Biological Effects*, Lyon, International Agency for Research on Cancer (*IARC Scientific Publications* No. 41) (in press)

3. PRINCIPLE

In a dry inert solvent the N-nitroso group in a compound is degraded to the amine by reaction with a solution of hydrobromic acid in glacial acetic acid.

4. HAZARDS

4.1 N-*Nitrosamine hazard*

N-Nitrosamines are carcinogenic and gloves must be worn during all operations involving handling of these compounds or their solutions. Moreover, N-nitrosamines in some solvents have been found to diffuse through many types of gloves (Gough et al., 1978; Walker et al., 1978; Sansone & Tewari, 1978).

Should the gloves come into contact with a solution of N-nitrosamine, they should be changed as quickly as possible to reduce the risk of contact of the N-nitrosamine with the skin.

It should also be borne in mind that many N-nitrosamines are volatile and so all operations should be carried out in a fume hood.

4.2 *Other hazards*

Hydrobromic acid/glacial acetic acid solutions are corrosive and should be handled with care.

5. REAGENTS[1]

5.1 *For decontamination*

Hydrobromic acid	30% solution in glacial acetic acid, Merck or equivalent
Hydrobromic acid solution	A solution containing at least 3% hydrogen bromide in glacial acetic acid. Prepared by dilution of 30% hydrobromic acid/glacial acetic acid with glacial acetic acid.

5.2 *For sample pretreatment or analysis*

Sodium hydroxide	10 mol/L solution (aqueous)
Dichloromethane	Technical grade
Sodium sulfate	Technical grade, anhydrous

6. APPARATUS

Usual laboratory equipment and the following items:

Gas chromatograph equipped with a suitable detection system

Magnetic stirrer

Ice bath

Kuderna-Danish concentrator (see 8.1 - notes on procedure)

Buchner funnel

7. PROCEDURE

A solution of *N*-nitrosamine in dichloromethane or other suitable solvent is concentrated, dried and treated with an excess of hydrobromic acid solution on the basis that 5 ml of the solution is sufficient to degrade 1 mg of *N*-nitrosamine in 1-2 ml of solvent within 15 min. NPYR is the exception to this requiring 10 ml of the hydrobromic acid solution to degrade 1 mg in 90 min.

The rate of reaction is drastically decreased by the presence of water or dimethyl sulfoxide (DMSO). In such a case Method 5 should be used.

---

[1] Reference to a company and/or product is for purposes of information and identification only and does not imply approval or recommendation of the company and/or product by the International Agency for Research on Cancer to the exclusion of others which may also be suitable.

7.1 *Undiluted N-nitrosamines*

7.1.1 Estimate the amount of N-nitrosamines to be destroyed and calculate the volume of hydrobromic acid solution required for their destruction.

7.1.2 Add twice the quantity of hydrobromic acid solution calculated in 7.1.1.

7.1.3 Allow to react at room temperature for at least 2 hr.

7.1.4 Check for completion of destruction (see 8.2 - notes on procedure)

7.1.5 Add excess of alkali to prevent re-formation of N-nitrosamines.

7.2 *Solutions in a volatile organic solvent*

7.2.1 Dry the solution by shaking with anhydrous sodium sulfate (see 7.7).

7.2.2 Filter the solution through Buchner funnel into a Kuderna-Danish evaporator and concentrate the filtrate to 1 to 2 ml.

7.2.3 Proceed as in 7.1.1 to 7.1.3.

7.2.4 See 8.2 if it is desired to check the residual solution for complete degradation.

7.2.5 Add excess alkali to prevent re-formation of N-nitrosamines.

7.3 *Solutions in ethanol*

The presence of ethanol drastically reduces the efficiency of hydrobromic acid denitrosation. Ethanol containing waste should be collected separately and treated as described below. In the presence of 1 ml of ethanol, 4 ml of hydrobromic acid solution will destroy 200 µg of N-nitrosamine in 24 hr. The content of ethanol in the reaction mixture should not exceed 20%.

7.3.1 Estimate the amount of N-nitrosamine to be destroyed and calculate the volume of hydrobromic acid solution required for its destruction.

7.3.2 Add a quantity of hydrobromic acid solution at least twice that calculated in 7.3.1 and ensure that the ethanol content is 20% or less.

7.3.3 Allow to react at room temperature for at least 24 hr.

7.3.4 Proceed as in 7.2.4 and 7.2.5.

7.4 *Aqueous solutions*

7.4.1 Extract with three successive volumes of dichloromethane; each volume being about half of the volume of water.

7.4.2 Proceed as in 7.2.1 to 7.2.5.

7.5 *Content of petri dishes*

NOTE: This method is not very suitable for dried agar.

7.5.1 Collect the contents of the petri dishes in a blender, add a volume of water equal to about twice the volume of the agar and homogenize.

7.5.2 Extract with three successive portions of dichloromethane, each portion being equivalent to about half the total amount of homogenate from 7.5.1. Combine the dichloromethane extracts.

7.5.3 Proceed as in 7.2.1 to 7.2.5.

7.6 *Spills of organic solutions (excluding DMSO solutions)*

7.6.1 Isolate the area, and put on suitable protective clothing, including breathing apparatus.

7.6.2 Absorb the spill with an absorbent material, such as blotting paper, place it immediately in a beaker and add a solution of hydrobromic acid in glacial acetic acid and leave overnight.

7.6.3 Place further absorbent material over the spill area and impregnate the area with hydrobromic acid solution. Close the area and leave overnight (see Egan et al., eds, 1982).

7.6.4 Check atmosphere for residual contamination by volatile N-nitrosamines.

7.6.5 Add solid sodium carbonate to the decontaminated surface.

7.6.6 Put on suitable protective clothing including breathing apparatus and remove the decontamination mixture and check the surface for completeness of decontamination.

7.7 *Solid wastes (e.g. column fillings, sodium sulfate etc.) and glassware*

7.7.1 Glassware: Drain thoroughly and treat solutions as described earlier.

7.7.2 Immerse glassware or solid waste in a solution of hydrobromic acid in glacial acetic acid and allow to react overnight. In the case of glassware polluted with an oily solution, allow to react for 48 hr.

7.7.3 Before disposal of the decontamination solution, ensure that no residual N-nitrosamines are present.

8. NOTES ON PROCEDURE

8.1 Alternative systems for concentration/evaporation may be used provided that they do not lead to loss of volatile N-nitrosamines.

8.2 *Check for residual* N-*nitrosamines after decontamination*

NOTE: This should be done by an analyst experienced in the analysis of N-nitrosamines using approved methods (see Egan et al., eds, 1978, 1982).

8.2.1 Transfer 5 ml of the media that has been subjected to decontamination process to a conical flask.

8.2.2 Cool in an ice bath. Add slowly, while stirring (preferably using a magnetic stirrer), 15 g of ice and 40 ml 10 M sodium hydroxide solution.

8.2.3 Transfer to a 100 ml separation funnel.

8.2.4 Extract with three successive 20 ml portions of dichloromethane.

8.2.5 Combine extracts and dry them over anhydrous sodium sulfate.

8.2.6 Concentrate extracts to about 1 ml using a Kuderna-Danish evaporator.

8.2.7 Analyse by gas chromatography. Method for analysis of nitrosamines other than N,N'-dinitrosopiperazine may be found in Egan et al., eds, 1978, 1982.
For analysis of N,N'-dinitrosopiperazine, the following conditions can be used:

Column:	Stainless steel 80 cm long × 1/8" (0.3175 cm) o.d. packed with 3% OV 17 on Chromosorb W, AW-DMCS, 80-100 mesh
Column temperature:	270°C
Injection temperature:	270°C
Carrier gas:	Argon ≃ 50 ml/min

METHOD I USING HYDROBROMIC ACID

9. SCHEMATIC REPRESENTATION OF PROCEDURE

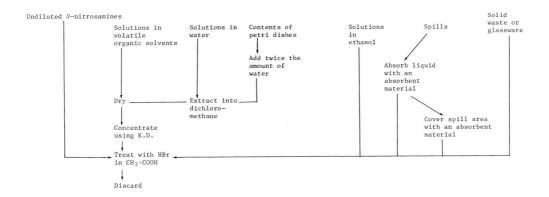

10. ORIGIN OF METHOD

International Agency for Research on Cancer
150 cours Albert Thomas
69372 Lyon Cédex 08, France

Contact point: Dr M. Castegnaro

METHOD 2
DESTRUCTION OF *N*-NITROSAMINES IN LABORATORY WASTES USING POTASSIUM PERMANGANATE

1. SCOPE AND FIELD OF APPLICATION

Procedures are specified for the treatment of undiluted *N*-nitrosamines (7.1), aqueous solutions (7.2) and spills (7.3).

The method has been tested for the following *N*-nitrosamines: *N*-nitrosodimethylamine (NDMA), *N*-nitrosodiethylamine (NDEA), *N*-nitrosodipropylamine (NDPA), *N*-nitrosodibutylamine (NDBA), *N*-nitrosopiperidine (NPIP), *N*-nitrosopyrrolidine (NPYR), *N*-nitrosomorpholine (NMOR) and *N*,*N*'-dinitrosopiperazine (Di-NPZ).

The method has been collaboratively studied using solutions of NDMA, NDBA and NPYR in water (50 mg/L, 25 mg/L and 50 mg/L, respectively).

The method affords better than 99.5% decontamination.

For recommended application of the method see Methods Index, p. 13.

2. REFERENCE

Castegnaro, M., Michelon, J. & Walker, E.A. (1982) In: Bartsch, H., O'Neill, I.K., Castegnaro, M. & Okada, M., eds, N-*Nitroso Compounds: Occurrence and Biological Effects*, Lyon, International Agency for Research on Cancer (*IARC Scientific Publications* No. 41) (in press)

3. PRINCIPLE

Decontamination is effected by oxidation with a solution of potassium permanganate in sulfuric acid.

4. HAZARDS

4.1 N-*Nitrosamine hazard*

N-Nitrosamines are carcinogenic and gloves must be worn during all operations involving handling of these compounds or their solutions. Moreover, *N*-nitrosamines in some solvents have been found to diffuse through many types of gloves (Gough et al., 1978; Walker et al., 1978; Sansone & Tewari, 1978). Should gloves come into contact with a solution of *N*-nitrosamines, they should be changed as quickly as possible to reduce the risk of contact of *N*-nitrosamines with the skin.

It should also be borne in mind that many *N*-nitrosamines are volatile and so all operations should be carried out in a fume hood.

4.2 *Other hazards*

Concentrated sulfuric acid is corrosive. Care should be taken in the preparation of solutions of potassium permanganate in sulfuric acid; never add solid potassium permanganate to concentrated sulfuric acid.

The dilution of concentrated sulfuric acid with water is extremely exothermic. Always add the acid to the water, never the reverse and remove heat by cooling in a cold water bath. If sulfuric acid is added to an aqueous solution containing volatile *N*-nitrosamines the flask must be fitted with a condenser to prevent *N*-nitrosamine volatilization.

5. REAGENTS

5.1 *For degradation*

Potassium permanganate	Technical grade
Sulfuric acid	Density 1.84 (about 18 mol/L)
Potassium permanganate/ sulfuric acid solution	Dilute concentrated sulfuric acid to 3 mol/L (see 4.2 Hazards) then add solid potassium permanganate to cool solution to obtain a 0.3 mol/L solution of potassium permanganate.

5.2 *For sample pretreatment or analysis*

Sodium hydroxide	10 mol/L solution (aqueous)
Dichloromethane	Technical grade
Sodium sulfate	Technical grade, anhydrous

6. APPARATUS

Usual laboratory equipment including items described under Method 1, section 6.

METHOD 2 USING POTASSIUM PERMANGANATE

7. PROCEDURE

   NOTE: 50 ml of the potassium permanganate/sulfuric acid reagent will degrade a mixture containing $\simeq$ 300 µg of N-nitrosamines. However, it should be noted that other components in the waste may react with potassium permanganate (turning the purple colour to brown). Thus in all cases sufficient permanganate should be added to maintain a permanent purple colour.

7.1 *Undiluted* N-*nitrosamines*

   7.1.1 Dissolve N-nitrosamine in 3 mol/L sulfuric acid solution

   7.1.2 Add sufficient potassium permanganate to obtain at least a 0.3 mol/L solution and also ensure that a purple colour remains after the reaction time (see 7.1.3).

   7.1.3 Leave to react, at room temperature, for at least 8 hr. A longer period of time is advisable, e.g. overnight.

   7.1.4 Check for completion of destruction (see Method 1, 8.2)

   7.1.5 Dilute with water

7.2 *Aqueous wastes*

   7.2.1 Add slowly, whilst stirring, enough sulfuric acid to the waste immersed in a cold water bath, to obtain a 3 mol/L solution (see 4.2 - Hazards).

   7.2.2 Proceed as in 7.1.2 to 7.1.3

   7.2.3 Dilute with water

7.3 *Spills of aqueous solutions*

   7.3.1 Isolate the area, and put on suitable protective clothing, including breathing apparatus, if considered necessary

   7.3.2 Absorb the spill with an absorbent material, such as blotting paper, place it immediately in a beaker and add a solution of potassium permanganate in sulfuric acid and leave overnight

   7.3.3 Pour some potassium permanganate/sulfuric acid solution over the contaminated area. Close the area and leave overnight.

   7.3.4 Check atmosphere for residual contamination by volatile N-nitrosamines (see Egan et al., eds, 1982)

   7.3.5 Add solid sodium carbonate to the decontaminated surface

   7.3.6 Remove the decontamination mixture by absorbing on an absorbent material. Check the surface and absorbent material for completeness of decontamination.

8. NOTES ON PROCEDURE

   See Method 1, 8.1

9. SCHEMATIC REPRESENTATION OF PROCEDURE

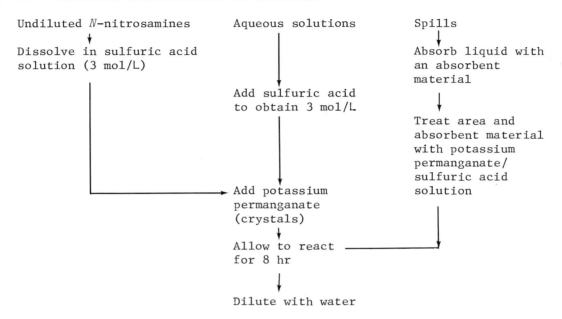

10. ORIGIN OF METHOD

    International Agency for Research on Cancer
    150 cours Albert Thomas
    69372 Lyon Cédex 08, France

    Contact point: Dr M. Castegnaro

METHOD 3
DECONTAMINATION OF *N*-NITROSAMINE CONTAMINATED GLASSWARE

1. SCOPE AND FIELD OF APPLICATION

This method specifies a procedure for the decontamination of glassware which contain *N*-nitrosamine solutions of concentrations not greater than 100 mg/L. For glassware contaminated with more concentrated solutions Method 1, 7.7, should be used.

The method has been collaboratively studied and affords effective decontamination, leaving ng amount or less on the glassware.

2. REFERENCE

None

3. PRINCIPLE

The glassware is rinsed five times with an appropriate solvent (see Methods Index, p. 13) and the combined rinses are decontaminated by a suitable method (see Methods Index, p. 13).

4. HAZARDS

4.1 N-*Nitrosamine hazard*

N-Nitrosamines are carcinogenic and gloves must be worn during all operations involving handling of these compounds or their solutions. Moreover, *N*-nitrosamines in some solvents have been found to diffuse through many types of gloves (Gough et al., 1978; Walker et al., 1978; Sansone & Tewari, 1978). Should gloves come into contact with a solution of *N*-nitrosamine, they should be changed as quickly as possible to reduce the risk of contact of *N*-nitrosamines with the skin.

It should also be borne in mind that many *N*-nitrosamines are volatile and so all operations should be carried out in a fume hood.

4.2 *Other hazards*

Hydrobromic acid and glacial acetic acid solutions are corrosive and should be handled with care.

5. REAGENTS[1]

Dichloromethane	Technical grade
Water	
Methanol	Technical grade
Hydrobromic acid	30% solution in glacial acetic Merck or equivalent
Hydrobromic acid solution	A solution containing at least 3% hydrogen bromide in glacial acetic acid prepared by dilution of 30% hydrobromic acid in glacial acetic acid with glacial acetic acid.

6. APPARATUS

Usual laboratory equipment including the following items:

Kuderna-Danish concentrator or equivalent

Alternative systems for concentration/evaporation may be used provided that they do not lead to loss of volatile N-nitrosamines (Egan et al., eds, 1978, 1982).

7. PROCEDURE

7.1 Rinse and drain the apparatus five times using an appropriate volume of solvent, and combine the rinses.

   NOTE: The volume used should be kept to a practical minimum bearing in mind that the solvent residues will themselves require decontamination.

7.2 Treat combined rinsings by an appropriate method (see Methods Index, p. 13)

8. NOTES ON PROCEDURE

None

---

[1] Reference to a company and/or product is for purposes of information and identification only and does not imply approval or recommendation of the company and/or product by the International Agency for Research on Cancer to the exclusion of others which may also be suitable.

9. SCHEMATIC REPRESENTATION OF PROCEDURE

    None

10. ORIGIN OF METHOD

    International Agency for Research on Cancer
    150 cours Albert Thomas
    69372 Lyon Cédex 08, France

    Contact point: Dr M. Castegnaro

METHOD 4
DECONTAMINATION OF LABORATORY WASTES USING TRIETHYLOXONIUM SALTS

1. SCOPE AND FIELD OF APPLICATION

This method specifies a procedure for the decontamination of laboratory wastes either dissolved in or extractable by dichloromethane.

The method has been collaboratively studied using triethyloxonium tetrafluoroborate for the treatment of solutions of N-nitrosodimethylamine, N-nitrosodibutylamine and N-nitrosopyrrolidine in dichloromethane (1 g/L each).

The method affords better than 99% decontamination.

For recommended application of the method see Methods Index, p. 13.

2. REFERENCES

Hünig, S., Geldern, L. & Lücke, E. (1963) *O-Alkyl-nitrosimmoniumsalze, eine neue Verbindungsklasse.* Angew. Chem., 75, 476

Meerwein, H., Hinz, G., Hofmann, P., Kroning, E. & Pfeil, E. (1937) Über tertiäre Oxoniumsalze, I. J. prakt. Chem., 147, 257-285

Meerwein, H., Battenberg, E., Gold, H., Pfeil, E. & Willfang, G. (1939) Über tertiäre Oxoniumsalze, II. J. prakt. Chem., 154, 83-156

Meerwein, H. (1966) Triethyloxonium Fluoroborate. Org. Synth., 46, 113-115

Castegnaro, M., Pignatelli, B. & Walker, E.A. (1974) *An investigation of the possible value of oxonium salt formation in nitrosamine analysis.* In: Bogovski, P. & Walker, E.A., eds, N-*Nitroso Compounds in the Environment*, Lyon, International Agency for Research on Cancer (*IARC Scientific Publications* No. 9), pp. 45-48

Castegnaro, M., Michelon, J. & Walker, E.A. (1982) In: Bartsch, H., O'Neill, I.K., Castegnaro, M. & Okada, M., eds, N-*Nitroso Compounds: Occurrence and Biological Effects*, Lyon, International Agency for Research on Cancer (*IARC Scientific Publications* No. 41) (in press)

3. PRINCIPLE

Formation of a non-volatile salt by the addition of an excess of triethyloxonium tetrafluoroborate to the dichloromethane extract, concentration in a rotary evaporator, and degradation using potassium hydroxide under reflux.

4. HAZARDS

4.1 N-*Nitrosamine hazard*

N-Nitrosamines are carcinogenic and gloves must be worn during all operations involving handling of these compounds or their solutions. Moreover, *N*-nitrosamines in some solvents have been found to diffuse through many types of gloves (Gough et al., 1978; Walker et al., 1978; Sansone & Tewari, 1978). Should gloves come into contact with a solution of *N*-nitrosamines, they should be changed as quickly as possible to reduce the risk of contact of *N*-nitrosamines with the skin.

It should also be borne in mind that many *N*-nitrosamines are volatile and so all operations should be carried out in a fume hood.

4.2 *Other hazards*

The reaction mixture must be considered to be potentially carcinogenic until it has been treated by strong alkali.

5. REAGENTS[1]

Triethyloxonium tetrafluoroborate	Solution in dichloromethane, Aldrich or equivalent
Sodium sulfate	Technical grade, anhydrous
Potassium hydroxide solution	10 mol/L (aqueous)

6. APPARATUS

Usual laboratory equipment including the following item:

Rotary evaporator

7. PROCEDURE

This procedure is applicable to all extracts in dichloromethane.

---

[1] Reference to a company and/or product is for purposes of information and identification only and does not imply approval or recommendation of the company and/or product by the International Agency for Research on Cancer, to the exclusion of others which may also be suitable.

METHOD 4 USING TRIETHYLOXONIUM SALTS

7.1 Dry the solution by shaking with anhydrous sodium sulfate

NOTE: For disposal of sodium sulfate see Method 1, 7.7.

7.2 Estimate the amount of N-nitrosamine present in the extract

7.3 Add an excess of triethyloxonium tetrafluoroborate (1 mole of TEOF reacts with 1 mole of N-nitrosamine)

7.4 Allow to react for a few minutes

7.5 Concentrate to a small volume using a rotary evaporator

7.6 Add 10 ml of a 10 mol/L potassium hydroxide solution and reflux for 1 hr.

7.7 If it is desired to check the residual solution for complete degradation analyse an aliquot of the dichloromethane (see Method 1, 8.2.4 to 8.2.7).

8. NOTES ON PROCEDURE

None

9. SCHEMATIC REPRESENTATION OF PROCEDURE

None

10. ORIGIN OF METHOD

International Agency for Research on Cancer
150 cours Albert Thomas
69372 Lyon Cédex 08, France

Contact point: Dr M. Castegnaro

METHOD 5
DESTRUCTION OF *N*-NITROSAMINES IN LABORATORY WASTES USING NICKEL-ALUMINIUM ALLOY IN POTASSIUM HYDROXIDE SOLUTION

1. SCOPE AND FIELD OF APPLICATION

This method specifies a procedure for the treatment of the following laboratory wastes contaminated with *N*-nitrosamines: undiluted nitrosamines (7.1), solutions in water (7.2), solutions in methanol (7.3), solutions in mineral oil (7.4), solutions in olive oil (7.5), solutions in dichloromethane (7.6), solutions in dimethyl sulfoxide (7.7), content of petri dishes (7.8). The *N*-nitrosamines investigated were *N*-nitrosodimethylamine (NDMA), *N*-nitrosodiethylamine (NDEA), *N*-nitrosodiisopropylamine (ND-i-PA), *N*-nitrosodibutylamine (NDBA), *N*-nitrosopiperidine (NPIP), *N*-nitrosopyrrolidine (NPYR), *N*-nitrosomorpholine (NMOR), *N*-nitrosomethylphenylamine (NMPA) and *N,N'*-dinitrosopiperazine (Di-NPZ).

The method affords better than 99.9% decontamination. The only product is the corresponding amine with < 0.1% of the corresponding hydrazine.

The method has been collaboratively studied using solutions of *N*-nitrosodimethylamine, *N*-nitrosodibutylamine and *N*-nitrosopyrrolidine in water (1 g/L each), olive oil (1 g/L each), dichloromethane (1 g/L each), DMSO (1 g/L each) and ethanol (1 g/L each) and for agar spiked with the same three *N*-nitrosamines at the levels of 4 mg, 4.8 mg and 4 mg, respectively.

The method affords better than 99.9% decontamination for all six substrates tested, provided that the powdered alloy has been kept in suspension, by stirring, throughout the reduction period (see Note 3 under 7. Procedure).

For recommended applications of the method see Methods Index, p. 13.

2. REFERENCE

Lunn, G., Sansone, E.B. & Keefer, L.K. (1981) Reductive destruction of *N*-nitrosodimethylamine as an approach to hazard control in the carcinogenesis laboratory. *Food Cosmet. Toxicol.*, *19*, 493-494

Sansone, E.B., Lunn, G., Jonas, L.A. & Keefer, L.K. (1982) *Approaches to hazard control in the carcinogenesis laboratory: N-nitroso compounds*. In: Bartsch, H., O'Neill, I.K., Castegnaro, M. & Okada, M., eds, N-*Nitroso Compounds: Occurrence and Biological Effects*, Lyon, International Agency for Research on Cancer (*IARC Scientific Publications* No. 41) (in press)

Seebach, D. & Wykypiel, W. (1979) Safe one-spot carbon.carbon bond formation with lithiated nitrosamines including denitrosation by sequential reduction with lithium aluminium hydride and Raney-nickel. *Synthesis*, 423-424

3. PRINCIPLE

The decontamination procedure consists of mixing the material to be decontaminated with a potassium hydroxide solution. Nickel-aluminium alloy is then added. Initially the aluminium reacts with the hydroxide ions and reduces the N-nitrosamine to the hydrazine. An active nickel powder which has adsorbed hydrogen is produced and this cleaves the hydrazine to the amine.

4. HAZARDS

4.1 N-*Nitrosamines*

N-Nitrosamines are carcinogenic and gloves must be worn during all operations involving handling of these compounds or their solutions. Moreover, N-nitrosamines in some solvents have been found to diffuse through many types of gloves (Gough et al., 1978; Walker et al., 1978; Sansone & Tewari, 1978). Should gloves come into contact with a solution of N-nitrosamine, they should be changed as quickly as possible to reduce the risk of contact of N-nitrosamines with the skin.

It should also be borne in mind that many N-nitrosamines are volatile and so all operations should be carried out in a fume hood.

4.2 *Other hazards*

Potassium hydroxide and its solutions are corrosive and appropriate protective clothing should be worn.

See also Notes in 7.1.3 and 7.1.5.

METHOD 5 USING NICKEL-ALUMINIUM AND KOH

5. REAGENTS[1]

5.1 *For degradation*

Potassium hydroxide solution	(approximately 1 mol/L) aqueous
Potassium hydroxide solution	(approximately 2 mol/L) aqueous
Nickel-aluminium alloy	50-50 powder, available from Aldrich Chemical Company (Gillingham, Dorset SP8 4BR, UK) and Alfa Ventron (Karlsruhe, FRG) or equivalent
Methanol	Technical product

5.2 *For sample pretreatment or analysis*

Dichloromethane	Technical grade
Sodium sulfate anhydrous	Technical grade
Celite	

6. APPARATUS

Usual chemical laboratory equipment and the following items:
Gas chromatograph equipped with a suitable detection system
Magnetic stirrer
Kuderna-Danish evaporator or any suitable concentration system
Büchner funnel

7. PROCEDURE

NOTE 1 : 50 g of nickel-aluminium alloy is sufficient to destroy 5 g of *N*-nitrosamines in 1 L of 0.5 mol/L potassium hydroxide solution. Other components in the waste may also react with the nickel-aluminium alloy or may poison the nickel catalyst which is formed. It is therefore recommended that the efficiency of the degradation be checked using the analytical procedure in section 8.

---

[1] Reference to a company and/or product if for purposes of information and identification only and does not imply approval or recommendation of the company and/or product by the International Agency for Research on Cancer, to the exclusion of others which may also be suitable.

NOTE 2 :   Addition of the nickel-aluminium alloy to an alkaline solution causes a highly exothermic reaction. If the reaction is uncontrolled, frothing can occur, distributing the solution all over the beaker. It is essential to add the Ni-Al alloy slowly over a period of time, while cooling the reaction vessel in an ice-bath.

NOTE 3 :   Care should be taken that the reductant alloy powder is kept in suspension throughout the whole operation (no formation of lumps or adhering to surfaces).

7.1 *Undiluted* N-*nitrosamines*

7.1.1 Take up 5 g of N-nitrosamines in 500 ml water (preferably) or methanol depending on the solubility of the N-nitrosamine (see Appendix A).

7.1.2 Commence stirring and dilute this solution with 500 ml of a 1 mol/L potassium hydroxide solution.

CAUTION: See introductory Note 2.

7.1.3 Whilst continuing to stir, cautiously add 50 g nickel-aluminium alloy per litre of N-nitrosamine solution over a period of 1 hour.

NOTE: If the addition of the alloy is too rapid, considerable foaming occurs and large quantities of hydrogen are evolved; therefore, the reaction should be performed in an efficient fume hood.

7.1.4 Continue to stir for about 24 hr.

7.1.5 Filter the mixture through a pad of Celite prepared by slurrying about 35 g of Celite in a 9 cm Büchner funnel.

NOTE: The solid may be pyrophoric when dry; therefore, it should not be allowed to dry in the presence of flammable solvents. The solid should be disposed of by transfering to a beaker and cautiously adding 1 mol/L hydrochloric acid and stirring until the reaction is complete; i.e., the black powder has dissolved. As an alternative, the Raney nickel generated by the method can be rendered nonpyrophoric by addition at the end of step 7.1.4 of 1 ml of acetone per gramme of alloy used and continuing stirring for at least 24 hours. At this step filtration can be performed as in 7.1.5.

METHOD 5 USING NICKEL-ALUMINIUM AND KOH

7.2 *Solutions in water*

CAUTION: See introductory Note 2.

7.2.1 If required, dilute the solution with water to a maximum content of N-nitrosamines of approximately 1% (See Appendix A for limits of solubility and note that these are given for pure water and that it will be probably less soluble in an alcaline solution).

7.2.2 Add an equal volume of a 1 mol/L potassium hydroxide solution and stir.

7.2.3 Proceed as in 7.1.3 to 7.1.5.

7.3 *Solutions in methanol*

CAUTION: See introductory Note 2.

7.3.1 If required, dilute the solution with methanol to a maximum content of N-nitrosamines of approximately 1%.

7.3.2 Proceed as in 7.2.2 and 7.2.3.

7.4 *Solutions in mineral oil*

CAUTION: See introductory Note 2.

7.4.1 Dilute with at least an equal volume of n-hexane or more if it is required to reduce the N-nitrosamine content to about 1%.

7.4.2 Add an equal volume of a 1 mol/L potassium hydroxide solution and stir.

7.4.3 Cover the reaction vessel to prevent excessive loss of hexane, but do not seal.

7.4.4 Proceed as in 7.1.3 to 7.1.5.

7.5 *Solutions in olive oil*

CAUTION: See introductory Note 2.

7.5.1 If required, dilute the solution with olive oil to a maximum content of N-nitrosamines of approximately 1%.

7.5.2 Proceed as in 7.2.2 to 7.2.3.

7.6 *Solutions in dichloromethane*

CAUTION: See introductory Note 2.

7.6.1 If required, dilute the solution with dichloromethane to a maximum content of N-nitrosamines of approximately 1%.

7.6.2 Begin stirring and add an equal volume of a 2 mol/L potassium hydroxide solution and 3 volumes of methanol (i.e., dichloromethane:water:methanol 1:1:3).

7.6.3 Proceed as in 7.1.3 to 7.1.5.

7.7 *Solutions in dimethyl sulfoxide*

CAUTION: See introductory Note 2.

7.7.1 If required, dilute the solution with methanol to a maximum content of N-nitrosamines of approximately 1%.

7.7.2 Proceed as in 7.2.2 and 7.2.3.

7.8 *Contents of petri dishes*

CAUTION: See introductory Note 2.

7.8.1 Lift the agar from the plates and transfer to a round-bottomed flask, together with 75 mL of a 1 mol/L potassium hydroxide solution per 17 g of agar.

7.8.2 Fit a reflux condenser to the flask and heat the mixture until all the agar is dissolved.

7.8.3 Allow to cool, stir with a magnetic stirrer and slowly add 2 g of nickel-aluminium alloy per 17 g of agar.

7.8.4 Proceed as in 7.1.4 and 7.1.5.

8. METHODS OF ANALYSIS

8.1 *For* N-*nitrosamines*

NOTE: The check for residual N-nitrosamines after decontamination should be performed by an analyst experienced in the analysis of N-nitrosamines using approved methods (Egan et al., eds, 1978, 1982).

Extract an aliquot from residue in 7.1.5 (or corresponding aqueous phase) with three successive equal volumes of dichloromethane. Combine extracts.

Dry extracts over sodium sulfate.

Concentrate using a Kuderna-Danish evaporator.

Analyse with a gas chromatograph equipped with a suitable sensitive detection system (e.g. nitrogen/phosphorus, N/P) or Thermal Energy Analyzer).

Gas-chromatographic column: Glass 1.8 long × 2 mm i.d. packed with 10% Carbowax 20 M + 2% KOH

METHOD 5 USING NICKEL-ALUMINIUM AND KOH

8.2 *For hydrazines*

The following gas-chromatographic conditions were used:

Column: Silane-treated glass 1.8 m long × 2 mm i.d. packed with 10% Carbowax 20 M + 2% KOH

Column temperature: 50-150°C depending on *N*-nitrosamine originally degraded

Carrier gas: Nitrogen 20-30 ml/min

9. SCHEMATIC REPRESENTATION OF PROCEDURE

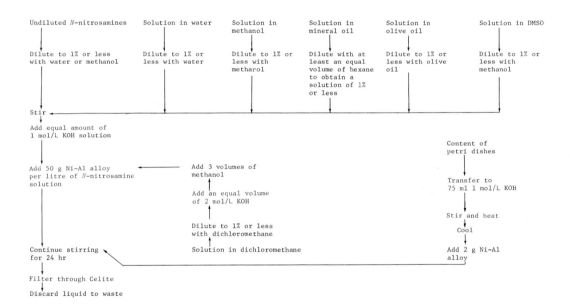

10. ORIGIN OF METHOD

NCI-Frederick Cancer Research Facility
P.O. Box B
Frederick, MD 21701, USA

Contact point: Dr E.B. Sansone

APPENDIX A
NOMENCLATURE AND CHEMICAL AND PHYSICAL DATA

1. *N*-NITROSAMINES

1.1 *N*-Nitrosodimethylamine

Nomenclature

Chemical Abstracts Registry Service No.: 62-75-9

Chemical Abstracts name: *N*-Methyl-*N*-nitrosomethanamine

Synonyms: Dimethylnitrosamine; *N,N*-dimethylnitrosamine; DMN; DMNA; NDMA; nitrosodimethylamine; *N*-dimethylnitrosamine; dimethyl-*N*-nitrosamine

Molecular and structural information

Molecular formula: $C_2H_6N_2O$

Molecular weight: 74.1

Structural formula: $O=N-N{<}^{CH_3}_{CH_3}$

Physical properties

Description: Yellow oily liquid (Magee & Barnes, 1967) Yellow oil which darkens in bright light (Hatt, 1943)

Boiling point: 151°C (760 mm) (Druckrey et al., 1967; Pensabene et al., 1972); 149-150°C (755 mm) (Hatt, 1943); 149-150°C (Looney et al., 1957); 146-147°C (704 mm) (Whitnack et al., 1963)

Density: d_4^{20} 1.0061 (IARC, 1978); d_4^{20} 1.0059 (CRC, 1976)

Refractive index: n_D^{20} 1.4368 (IARC, 1978); n_D^{20} 1.4395 (Heyns & Roper, 1971) n_D^{18} 1.4374 (Looney et al., 1957) n_D^{20} 1.4358 (CRC, 1976)

Solubility: Soluble in water, organic solvents and lipids (IARC, 1978; Druckrey et al., 1967)

Spectral properties: (see Table 2)

1.2 N-Nitrosodiethylamine

Nomenclature

Chemical Abstracts Registry Service No.: 55-18-5

Chemical Abstracts name: N-Ethyl-N-nitrosoethanamine

Synonyms: N,N-Diethylnitrosamine; diethylnitrosamine; nitrosodiethylamine; DEN; DENA; NDEA; N-diethylnitrosamine; diethyl-N-nitrosamine

Molecular and structural information

Molecular formula: $C_4H_{10}N_2O$

Molecular weight: 102.1

Structural formula: $O=N-N\begin{smallmatrix}CH_2-CH_3\\CH_2-CH_3\end{smallmatrix}$

Physical properties

Description: Yellow volatile liquid (IARC, 1978)

Boiling point: 177°C (Heyns & Roper, 1971; Looney et al., 1957); 64-65°C (17 mm) (Druckrey et al., 1967); 121-122°C (140 mm) (Whitnack et al., 1963); 64°C (17 mm) (Pensabene et al., 1972)

Density: d_4^{20} 0.9422 (IARC, 1978; CRC, 1976)

Refractive index: n_D^{20} 1.4386 (Looney et al., 1957; CRC, 1976); n_D^{20} 1.4410 (Heyns & Roper, 1971)

Solubility: Soluble in water: 10.6 g/100 ml (Druckrey et al., 1967) Soluble in organic solvents and lipids

Spectral properties: (see Table 2)

1.3 *N*-Nitrosodipropylamine

Nomenclature

Chemical Abstracts Registry Service No.: 621-64-7

Chemical Abstracts name: *N*-Nitroso-*N*-propyl-1-propanamine

Synonyms: *N*,*N*-di-*n*-propylnitrosamine; DPNA; NDPA; dipropyl-nitrosamine; di-*n*-propylnitrosamine; di-*n*-propylnitrosoamine (Although no other examples were found, as a rule, the term nitrosoamine would be applicable to other dialkyl derivatives)

Molecular and structural information

Molecular formula: $C_6H_{14}N_2O$

Molecular weight: 130.2

Structural formula: $O=N-N\begin{smallmatrix}CH_2-CH_2-CH_3\\CH_2-CH_2-CH_3\end{smallmatrix}$

Physical properties

Description: Yellow liquid (IARC, 1978)

Boiling point: 101-104°C (25 mm) (Heyns & Roper, 1971); 81°C (5 mm) (Druckrey et al., 1967; Pensabene et al., 1972); 90°C (13 mm) (Pulidori et al., 1970); 80-82°C (2 mm) (Whitnack et al., 1963)

Density: d_4^{20} 0.9160 (IARC, 1978); d_4^{25} 0.91101 (Pulidori et al., 1970); d_4^{20} 0.9163 (CRC, 1976)

Refractive index: n_D^{20} 1.4487 (Heyns & Roper, 1971); n_D^{20} 1.4437 (IARC, 1978; CRC, 1976); n_D^{25} 1.44165 (Pulidori et al., 1970)

Solubility: Soluble in water: 0.98 g/100 ml (Druckrey et al., 1967); Soluble in organic solvents and lipids

Spectral properties: (see Table 2)

1.4 N-Nitrosodibutylamine

Nomenclature

Chemical Abstracts Registry Service No.: 924-16-3

Chemical Abstracts name: N-Butyl-N-nitroso-1-butanamine

Synonyms: N,N-Di-n-butylnitrosamine; DBNA; DBN; NDBA; dibutylnitrosamine; N-dibutylnitrosamine; N,N-dibutylnitrosamine; N-nitrosodi-n-butylamine; dibutyl-N-nitrosamine

Molecular and structural information

Molecular formula: $C_8H_{18}N_2O$

Molecular weight: 158.2

Structural formula: $O=N-N\begin{subarray}{l}CH_2-CH_2-CH_2-CH_3\\CH_2-CH_2-CH_2-CH_3\end{subarray}$

Physical properties

Description: Yellow oil (IARC, 1978)

Boiling point: 116°C (14 mm) (Druckrey et al., 1967; Pensabene et al., 1972); 115°C (17 mm) (Heyns & Roper, 1971); 133°C (130 mm) (Looney et al., 1957)

Density: d_4^{20} 0.9009 (IARC, 1978)

Refractive index: n_D^{20} 1.4510 (Heyns & Roper, 1971); n_D^{20} 1.4475 (IARC, 1978); n_D^{24} 1.4447 (Looney et al., 1957)

Solubility: Soluble in water: 0.12 g/100 ml (Druckrey et al., 1967); Soluble in organic solvents and vegetable oils

Spectral properties: (see Table 2)

1.5 N-Nitrosopiperidine

Nomenclature

Chemical Abstracts Registry Service No.: 100-75-4

Chemical Abstracts name: 1-Nitroso-piperidine

Synonyms: NO-Pip; NPip; NPIP; nitrosopiperidine; N-piperidyl nitrosamine

Molecular and structural information

Molecular formula: $C_5H_{10}N_2O$

Molecular weight: 114.2

Structural formula:

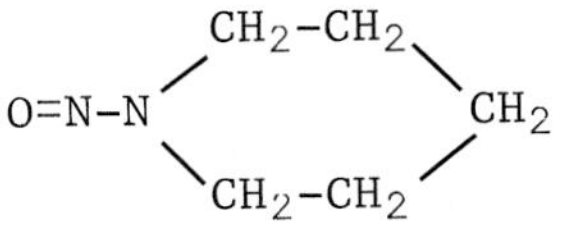

$$O=N-N\begin{array}{c}CH_2-CH_2\\ \\CH_2-CH_2\end{array}CH_2$$

Physical properties

Description: Yellow oil (IARC, 1978)

Boiling point: 215°C (721 mm) (IARC, 1978); 100°C (14 mm) (Druckrey et al., 1967; Pensabene et al., 1972); 110°C (17 mm); 102°C (13 mm) (Heyns & Roper, 1971); 85°C (0.5 mm) (Jones & Kenner, 1932); 217°C (721 mm); 109°C (20 mm) (CRC, 1976)

Density: $d_4^{18.5}$ 1.0631 (CRC, 1976)

Refractive index: $n_D^{18.5}$ 1.4933 (IARC, 1973; CRC, 1976); n_D^{20} 1.4958; n_D^{20} 1.4950 (Heyns & Roper, 1971)

Solubility: Soluble in water: 7.7 g/100 ml (Druckrey et al., 1967); soluble in organic solvents and lipids

Spectral properties: (see Table 2)

1.6 *N*-Nitrosopyrrolidine

Nomenclature

Chemical Abstracts Registry Service No.: 930-55-2

Chemical Abstracts name: 1-Nitrosopyrrolidine

Synonyms: *N*-Pyr; NO-Pyr; NPYR; nitrosopyrrolidine; NPy

Molecular and structural information

Molecular formula: $C_4H_8N_2O$

Molecular weight: 100.2

Structural formula:

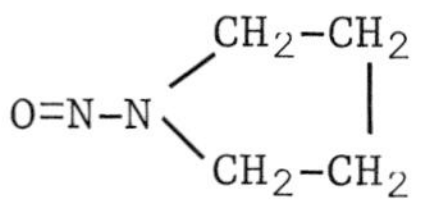

$$O=N-N\begin{matrix} CH_2-CH_2 \\ | \\ CH_2-CH_2 \end{matrix}$$

Physical properties

Description: Yellow liquid (IARC, 1978)

Boiling point: 98°C (12 mm) (Druckrey et al., 1967; Pensabene et al., 1972); 214°C (760 mm) (IARC, 1978)

Density: 1.10 (Groenen et al., 1976)

Refractive index: No data available

Solubility: Miscible with water in all proportions (Druckrey et al., 1967); soluble in organic solvents and lipids

Spectral properties (see Table 2)

1.7 N-Nitrosomorpholine

Nomenclature

Chemical Abstracts Registry Service No.: 59-89-2

Chemical Abstracts name: 4-Nitrosomorpholine

Synonyms: NMOR; nitrosomorpholine

Molecular and structural information

Molecular formula: $C_4H_8N_2O_2$

Molecular weight: 116.1

Structural formula:

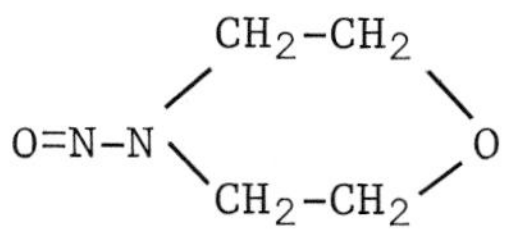

$$O=N-N\begin{matrix} CH_2-CH_2 \\ CH_2-CH_2 \end{matrix}O$$

Physical properties

Description: Yellow crystals (IARC, 1978)

Boiling point: 96°C (6 mm) (Druckrey et al., 1967; Pensabene et al., 1972); 224-225°C (747 mm) (IARC, 1978); 101°C (17 mm) (Heyns & Roper, 1970); 98°C (10 mm) (Looney et al., 1957)

Melting point: 29°C (Looney et al., 1957)

Density: No data available

Refractive index: No data available

Solubility: Miscible with water in all proportions (Druckrey et al., 1967)

Spectral properties: (see Table 2)

1.8 *N,N'*-Dinitrosopiperazine

Nomenclature

Chemical Abstracts Registry Service No.: 140-79-4

Chemical Abstracts name: 1,4-Dinitrosopiperazine

Synonyms: Dinitrosopiperazine; 1,4-dinitrosopiperazine; DNP

Molecular and structural information

Molecular formula: $C_4H_8N_4O_2$

Molecular weight: 144.1

Structural formula:

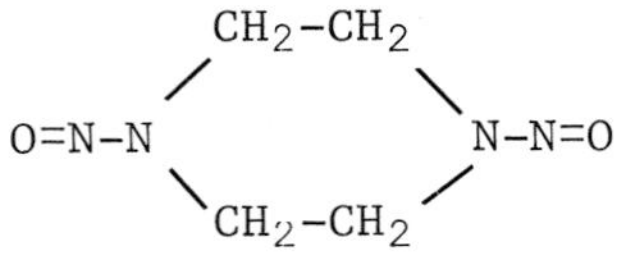

$$O=N-N\begin{matrix}\diagup CH_2-CH_2 \diagdown \\ \diagdown CH_2-CH_2 \diagup\end{matrix}N-N=O$$

Physical properties

Description: No data available

Boiling point: 160°C (Druckrey et al., 1967); 155-156°C (Lijinsky & Taylor, 1975); 159-161°C (Krüger et al., 1976)

Density: No data available

Refractive index: No data available

Solubility: Soluble in water: 0.57 g/100 ml

Spectral properties: (see Table 2)

2. SPECTRAL PROPERTIES

NMR and infrared spectra of some of the N-nitrosamines (NDMA, NDEA, NDBA and NMOR) have been studied by Looney et al., 1957 and reviewed by Rao and Bhaskar, 1969. Mass spectra are presented by Pensabene et al. (1972) and Rainay et al. (1978), and the fragmentation patterns discussed. Ultraviolet absorption data are presented in Table 2., the dependence of the maxima on solvent has been thoroughly studied by Haszeldine and Mattison (1955).

Table 2

N-Nitrosamine	Ultraviolet absorption maxima			Reference
	Solvent	nm	ε	
N-Nitrosodimethyl-amine	Light petroleum	232 351 361 374	5 900 98 125 105	Haszeldine & Jander (1954)
	Ethanol	231 346	7 000 100	Haszeldine & Jander (1954)
	Water	230 332	7 244 95	Druckrey et al. (1967)
N-Nitrosodiethyl-amine	Light petroleum	233 366 378	6 500 105 90	Haszeldine & Jander (1954)
	Ethanol	233 350	7 400 90	Haszeldine & Jander (1954)
	Water	230 340	7 413 85	Druckrey et al. (1967)
N-Nitrosodipropyl-amine	Water	233 339	7 585 85	Druckrey et al. (1967)
N-Nitrosodibutyl-amine	Water	233 347	7 079 89	Druckrey et al. (1967)
N-Nitrosopyrrolidine	Water	230 333	8 128 107	Druckrey et al. (1967)

Table 2 (cont'd)

N-Nitrosamine	Ultraviolet absorption maxima			Reference
	Solvent	nm	ε	
N-Nitrosopiperidine	Light petroleum	238 365 377	4 200 70 65	Haszeldine & Jander (1954)
	Ethanol	235 351	8 100 95	Haszeldine & Jander (1954)
	Water	235 337	9 549 83	Druckrey et al. (1967)
N-Nitrosomorpholine	Water	237 346	7 943 85	Druckrey et al. (1967)
N,N'-Dinitroso-piperazine	Water	239 337	15 135 104	Druckrey et al. (1967)
	Water	341	178	Lijinsky & Taylor (1975)
	Methanol	241 355	2 470 324	Krüger et al. (1976)

APPENDIX B
FURTHER STUDIES RELEVANT TO THE DEGRADATION OF *N*-NITROSAMINES

1. BIOLOGICAL DEGRADATION

Degradation of *N*-nitrosamines by biological routes shows low efficiency. As can be seen from Table 3, only a maximum of 50% has been observed in sewage (Tate & Alexander, 1975).

Table 3. Biological degradation of *N*-nitrosamines

N-Nitrosamines investigated	Type of biological degradation	Efficiency	Reference
NDMA, NDEA and NDPA	*P. statzeri, P. fragi, B. mycoides, B. cereus, B. subtilis, C. suboxydans, C. pasteurianum, A. picolinophilus, A. suboxydans* and 4 unclassified strains of arthrobacter were incubated for 72 hr with *N*-nitrosamines	No evidence of degradation	Tate & Alexander (1976)
NDMA, NDEA and NDPA	Natural disappearance in lake water, soil and sewage	Lake water: persistant over 3-5 months. Soil: slow disappearance over several weeks. Sewage: 50% drop in 14 days	Tate & Alexander (1975)
NDPA	Natural disappearance in soil under anaerobic or aerobic conditions	23% drop in 60 days under anaerobic conditions	Saunders et al. (1979)
NDMA, NDEA, NDPA, NPYR and NPIP	Degradation by: *Rhizopus orizae, Streptococcus cremoris, Saccharomyces rouxii*	80% drop for NDPA, 40% for NPIP, 25% for NDEA, 20% for NPYR and 10% for NDMA by *Rhizopus orizae*. Lower efficiency for other strains.	Katsuhiko & Kinjiro (1979)

2. CHEMICAL REACTIONS

Chemical reactions of N-nitrosamines have been reported in the literature some of which can be useful in establishing methods of degradation. Reviews on the chemical properties of N-nitrosamines have been published by Leotte (1964), Smith (1966), Fridman et al. (1971) and Anselme (1979).

Table 4 summarizes the available information on these reactions together with the percentages of conversion, when they are reported, and the source of the reference. The available information on toxicity of some of the products is discussed on page 5.

Table 4. Chemical reactions of N-nitrosamines

4.1 Oxidation

N-Nitrosamines	End products	Reaction and % conversion	Reference
NDMA, NDEA, NDPA and NDBA	Corresponding nitramines	Peroxotrifluoroacetic acid oxidizes nitrosamine in dichloromethane Efficiency 73-95% A trifluoroacetic acid/hydrogen peroxide (90%) solution achieves a conversion of around 75%, except for NPIP 86%	Emmons W.D. (1954)
NDMA	Nitrodimethylamine	A trifluoroacetic acid/hydrogen peroxide (50%) solution (5:4) oxidizes NDMA in solution, in dichloromethane in 12-24 hr Efficiency 50-60%	Sen (1970)
NDMA, NDEA, NDPA, NDBA, NPIP and NPYR	Corresponding nitramines	Peroxotrifluoroacetic acid oxidizes nitrosamines in dichloromethane solution in 3 hr Efficiency around 85%, except for NPIP where varying levels of nitraminopiperidine are detected. NPIP is completely degraded under these conditions	Althorpe et al. (1970) Telling (1972)
NDMA, NDEA, NDPA, NDBA, NPIP and NPYR	Corresponding nitramines	Comparison of efficiency of oxidation of diluted nitrosamine solutions using hydrogen peroxide 50% or 85%, and trifluoracetic acid with 50% hydrogen peroxide: 80-90% yields can be achieved except NDMA 60% and NPIP 70%	Castegnaro & Walker (1976)
NDEA, NDBA, NPIP and NMOR	Nitrogen, ammonia, nitric acid	Oxidation with concentrated sulfuric acid at high temperature	Kozak et al. (1972)
NDMA		γ-Radiolysis in aqueous solution causes oxidation. This is promoted in acid solution and prevented by addition of cysteine glutathione and ascorbic acid	Hirano (1973)

Table 4. (cont'd)

4.2 Photolysis

N-Nitrosamines	End products	Reaction and % conversion	Reference
NDMA, NDEA NDBA and NPIP		The photolytic efficiency is dependent on the nature of the solvent and on the structure of the nitrosamine	Gowenlock et al. (1978); Dent & Burns (1970)
NDMA, NDEA, NPYR, NPIP and NMOR		Photolysis at 366 nm in a capillary tube destroys nitrosamines at 0.5 ng/μL level. In dichloromethane or hexane completely in 60 min In methanol 98-100% in 90 min In water 82-93% in 90 min	Doerr & Fiddler (1977)
NDPA	Nitrite	Photolysis in a 5 mM NaOH solution using a xenon arc lamp in a continuous flow cell at 0.3 ml/min Efficiency 97%	Snider & Johnson (1979)
NDMA, NDEA, NDPA, NDBA and NPIP	Nitrite	Irradiation of paper or TLC chromatograms by direct sunlight	Vasundhara et al. (1975)
NDMA	Nitric oxide, carbon monoxide, formaldehyde + unidentified	Exposure in the vapour phase to direct sunlight for about 30 min Efficiency 50%	Hanst et al. (1977)
NDMA, NDPA and NMOR	Nitrite	UV irradiation of solutions in 0.5% Na_2CO_3 Efficiency > 95%	Daiber & Preussmann (1964)
NDBA	Butyraldehyde N-butylbutramidoxime dibutylamine	Irradiation of a 0.2 N HCl solution in methanol:water (1:1) with a "Hanovia" lamp	Chow (1964)
NPIP	Tetrahydropyridine, 2-piperidonoxime		
NDBA and NPIP		Prolonged irradiation under nitrogen of solutions in solvents such as water, alcohol, cyclohexane provokes no appreciable degradation. Addition of acid before irradiation brings a quick disappearance of the UV spectrum	Chow (1967)

Table 4. (cont'd)

4.4 Other reactions

N-Nitrosamines	End products	Reaction and % conversion	Reference
NDMA	Dimethylamino-pyrimidine + 5-nitroso pyrimidines	Heat at 175-190°C for 3 hr with chloropyrimidine and a 5-activated pyrimidine	Yoneda et al. (1973)
NDEA	Various fluorinated products	Electro-fluorination: variation of the conditions leads to different products	Watanabe et al. (1972)
N-Nitrosamines		Formation of a salt with triethyloxonium tetrafluoroborate which dissolves without decomposition in hydroxylic solvents and evolves nitrogen in strong bases	Hünig et al. (1963)
NDMA, NDEA, NDPA, NDBA, NPIP and NPYR		Formation of a salt with triethyloxonium tetrafluoroborate in dichloromethane solution which can be oxidized to nitramines and hydrolysed in boiling water	Castegnaro et al. (1974)
N-Nitrosamines		Reacts readily with trialkyloxonium hexachloroantimonate and at 60°C under nitrogen with methyl iodide	Schmidpeter (1963,1964)
NDMA		Reacts slowly with thiols under nitrogen but the mechanism is not clarified	Waters (1978)
NDMA, NDEA, NDPA, NDBA, NPIP and NPYR		Treatment with heptafluorobutyric anhydride in pyridine - mass spectra of the derivatives are given with a tentative structure	Gough et al. (1975,1976)
NDMA, NDEA, NPIP and NPYR		Treatment with heptafluorobutyric anhydride in pyridine for 1 hr	Brooks et al. (1972)
NDMA and NPYR		Thermal treatment at 110°C at various pH Best efficiency for NPYR 50% in 1 day at pH 12.2 for NDMA 50% in 67 days at pH 12.2	Fan & Tannenbaum (1972)
NPIP		Thermal decomposition in the range 265-312°C: homogeneous non-molecular process	Golovanova et al. (1974)

Table 4. (cont'd)

4.4 Other reactions (continued)

N-Nitrosamines	End products	Reaction and % conversion	Reference
NDMA, NDEA, NDPA, NDBA, NPIP and NPYR		Stability to heat under neutral, alkaline (0.2N NaOH) or acid (0.2N tartaric acid) conditions	Eisenbrand et al. (1970)
N-Nitroso C-Nitroso N-Nitro compounds		Compounds containing a N-N bond burn 2 to 4 times more quickly than compounds having C-N bonds	Fogel'Zang et al. (1974)
Volatile nitrosamines		Pyrolysis of nitrosamines in the gas phase at 550°C	Von Rappard et al. (1976)
NDMA	Dimethylamine	Action of a 1.1 equivalent of phenyl copper at -80°C for 3 hr Efficiency about 80%	Rahman et al. (1976)

REFERENCES

Alliston, T.G., Cox, G.B. & Kirk, R.S. (1972) The determination of steam-volatile nitrosamines in foodstuffs by formation of electron-capturing derivatives from electrochemically derived amines. *Analyst (Lond.), 97*, 915-920

Althorpe, J., Goddard, D.A., Sissons, D.J. & Telling, G.M. (1970) The gas chromatographic determination of nitrosamines at the picogram level by conversion to their corresponding nitramines. *J. Chromat., 53*, 371-373

Anselme, J.P. (1979) N-*Nitrosamines*, A.C.S. Symp. Ser. No. 101, American Chemical Society, Washington D.C.

Backer, M.H.J. (1913) Réductions électrochimiques: réduction des nitrosamines. *Rec. Trav. chim. Pays-Bas Belg., 32*, 39-47

Bamford, C.H. (1939) A study of the photolysis of organic nitrogen compounds. Part 1: Dimethyl- and diethylnitrosamines. *J. Chem. Soc.*, 12-17

Bartsch, H., Margison, G.P., Malaveille, C., Camus, A.M., Brun, G. & Margison, J.M. (1977) Some aspects of metabolic activation of chemical carcinogens in relation to their organ specificity. *Arch. Toxicol., 39*, 51-63

Bighi, C., Gilli, G. & Pulidori, F. (1968) Momento electrici dipolarie struttura di N-Nitrosoammine alifatiche. *Annali Univ. Ferrara, Sezione 5, 2*, 205-226

Bogovski, P. & Bogovski, S. (1981) Animal species in which N-nitroso compounds induce cancer. *Int. J. Cancer, 27*, 471-474

Brooks, J.B., Alley, C.C. & Jones, R. (1972) Reaction of nitrosamine with fluorinated anhydrides and pyridine to form electron capturing derivatives. *Analyt. Chem., 44*, 1881-1884

Burgess, E.M. & Lavanish, J.M. (1964) Photochemical decomposition of N-nitrosamines. *Tetrahedron Lett.*, 1221-1226

Castegnaro, M. & Walker, E.A. (1976) *Developments in nitrosamine analysis*. In: Tinbergen, B.J. & Kroll, B., eds, *Proc. 2nd Int. Symp. Nitrite Meat Prod.*, Zeist, Wageningen, PUDOC, The Netherlands, pp. 187-190

Castegnaro, M. & Walker, E.A. (1978) *New data from collaborative studies on analysis of nitrosamines.* In: Walker, E.A., Castegnaro, M., Griciute, L. & Lyle, R.E., eds, *Environmental Aspects of N-Nitroso Compounds*, Lyon, International Agency for Research on Cancer (*IARC Scientific Publications* No. 19), pp. 53-62

Castegnaro, M. & Walker, E.A. (1980) *Report on collaborative studies on determination of volatile nitrosamines in cheese and pesticides.* In: Walker, E.A., Griciute, L., Castegnaro, M. & Börszönyi, M., eds, *N-Nitroso Compounds: Analysis, Formation and Occurrence*, Lyon, International Agency for Research on Cancer (*IARC Scientific Publications* No. 31), pp. 445-453

Castegnaro, M., Pignatelli, B. & Walker, E.A. (1974) *An investigation of the possible value of oxonium salt formation in nitrosamine analysis.* In: Bogovski, P. & Walker, E.A., eds, *N-Nitroso Compounds: Analysis and Formation*, Lyon, International Agency for Research on Cancer (*IARC Scientific Publications* No. 9), pp. 45-48

Cessna, A.J., Sugamori, S.E., Yip, R.W., Lau, M.P., Snyder, R.S. & Chow, Y.L. (1977) Flash photolysis studies of N-chloro- and N-nitrosopiperidine. Assignement and reactivity of the piperidinium radical. *J. Am. Chem. Soc., 99*, 4044-4048

Chien, P.T. & Thomas, M.H. (1978) Evaluation of a degradation method for nitrosamine wastes. *J. Environ. Pathol. Toxicol., 2*, 513-516

Chow, Y.L. (1964) Photolysis of N-nitrosamines. *Tetrahedron Lett.*, 2333-2338

Chow, Y.L. (1967) Photochemistry of nitroso compounds in solution. V. Photolysis of N-nitrosodialkylamines. *Can. J. Chem., 45*, 53-62

Cox, G.B. (1973) Estimation of volatile N-nitrosamines by high performance liquid chromatography. *J. Chromat., 83*, 471-481

Chemical Rubber Company (1976) *Handbook of Chemistry and Physics*, 54th edition, Weast, R.C., ed., Cleveland, Ohio

Crosby, D.G., Humphrey, J.R. & Moilanen, K.W. (1980) The photodecomposition of dipropylnitrosamine vapor. *Chemosphere, 9*, 51-54

Crosby, N.T. & Sawyer, R. (1976) N-Nitrosamines: a review of chemical and biological properties and their estimation in foodstuffs. *Adv. Food Res., 22*, 1-71

Daiber, D. & Preussmann, R. (1964) Quantitative colorimetrische Bestimmung organischer N-Nitroso Verbindungen durch photochemische Spaltung der Nitrosaminbindung. *Z. analyt. Chem., 206*, 344-355

Dent, G.V. & Burns, D.T. (1970) Some studies on the decomposition stage in the estimation of N-Nitrosamines. Loughborough University of Technology, Department of Chemistry. *Summaries of Final Year Student Project Theses, 11*, 18-21

REFERENCES

Derr, P.F. (1960) Process for the chemical reduction of nitrosamines. *US Patent No. 2 961 467*

Doerr, R.C. & Fiddler, W. (1977) Photolysis of volatile nitrosamines at picogram level as an aid to confirmation. *J. Chromat., 140*, 284-287

Douglass, M.L., Kabacoff, B.L., Anderson, G.A. & Cheng, M.C. (1978) The chemistry of nitrosamine formation, inhibition and destruction. *J. Soc. cosmet. Chem., 29*, 581-606

Drescher, G.S. & Franck, C.W. (1978) Estimation of extractable N-nitroso compounds at the part per billion level. *Analyt. Chem., 50*, 2118-2121

Druckrey, H., Preussmann, R., Ivankovic, S. & Schmahl, D. (1967) Organotrope carcinogen Wirkungen bei 65 verschiedenen N-nitroso Verbindungen an BD-Ratten. *Z. Krebsforsch., 69*, 103-201

Egan, H., ed.-in-c., Preussmann, R., Castegnaro, M., Walker, E.A. & Wasserman, A.E., eds, (1978) *Environmental Carcinogens. Selected Methods of Analysis*, Vol. 1, *Analysis of Volatile Nitrosamines in Food*, Lyon, International Agency for Research on Cancer (*IARC Scientific Publications* No. 18)

Egan, H., ed.-in-c., Preussmann, R., O'Neill, I.K., Eisenbrand, G., Spiegelhalder, B., Castegnaro, M. & Bartsch, H., eds, (1982) *Environmental Carcinogens. Selected Methods of Analysis*, Vol. 6, N-*Nitroso Compounds*, Lyon, International Agency for Research on Cancer (*IARC Scientific Publications* No. 45)

Eisenbrand, G. & Preussmann, R. (1970) Eine neue Methode zur Kolorimetrischen Bestimmung von Nitrosaminen nach Spaltung der Nitrosogruppe mit Bromwasserstoff in Eisessig. *Arzneim.-Forsch., 20*, 1513-1517

Eisenbrand, G., Von Hodenberg, A. & Preussmann, R. (1970) Trace analysis of N-nitroso compounds. II. Steam distillation at neutral, alkaline and acid pH under reduced and atmospheric pressure. *Z. analyt. Chem., 251*, 22-24

Eizember, R.F., Vogler, K.R., Souter, R.W., Cannon, W.N. & Wege, II, P.M. (1979) Destruction of nitrosamines. Treatment of nitrosamines with various acids and halogens. *J. org. Chem., 44*, 784-786

Emmett, G.C., Michejda, C.J., Sansone, E.B. & Keefer, L.K. (1979) Limitation of photodegradation in the decontamination and disposal of chemical carcinogens. In: Walters, D.B., ed., *Safe Handling of Chemical Carcinogens, Mutagens, Teratogens and Highly Toxic Substances*, Vol. 2, Ann Arbor Science, Ann Arbor, MI, USA 1980, pp. 535-553

Emmons, W.D. (1954) Peroxotrifluoracetic acid. I. The oxidation of nitrosamines to nitramines. *J. Am. chem. Soc., 76*, 3468-3470

Fajen, J.M., Carson, G.A., Rounbehler, D.P., Fan, T.Y., Vita, R., Goff, U.E., Wolf, M.H., Edwards, G.S., Fine, D.H., Reinhold, V. & Biemann, K. (1979) *N*-Nitrosamines in the rubber and tyre industry. *Science, N.Y., 205*, 1262-1264

Fajen, J.M., Rounbehler, D.P. & Fine, D.H. (1982) *Summary report on N-nitrosamines in the factory environment*. In: Bartsch, H., O'Neill, I.K., Castegnaro, M. & Okada, M., eds, N-*Nitroso Compounds: Occurrence and Biological Effects*, Lyon, International Agency for Research on Cancer (*IARC Scientific Publications* No. 41) (in press)

Fan, T.S. & Tannenbaum, S.S. (1971) Automatic colorimetric determination of *N*-nitroso compounds. *J. Agric. Food Chem., 19*, 1267-1269

Fan, T.S. & Tannenbaum, S.R. (1972) Stability of *N*-nitroso compounds. *J. Food Sci., 37*, 274-276

Farina, P.R. (1970) Hexahydrotetrazines from nitrosamines. *Tetrahedron Lett.*, 4971-4973

Farina, P.R. (1972) The reaction of organometallics with nitrosamines and the nitrosylation of folic acid and related compounds. *Diss. Abstr. Int. B., 33B*, 1450

Farina, P.R. & Tieckelmann, H. (1973) Some reactions of organolithium compounds with nitrosamines. *J. org. Chem., 38*, 4259-4263

Farina, P.R. & Tieckelmann, H. (1975) Reactions of Grignard reagents with nitrosamines. *J. org. Chem., 40*, 1070-1074

Fogel'Zang, A.E., Svetlov, B.S., Adzhemyan, V. Ya., Kolyasov, S.M. & Sergienko, O.I. (1974) The combustion of nitramines and nitrosamines. *Dok. Akad. Nauk SSSR, 216*, 603-606

Franc, J. & Mikës, F. (1967) Systematic analysis of nitrogen compounds using cleavage reactions and gas chromatography. *J. Chromat., 26*, 378-386

Fridman, A.L., Mukhametshin, F.M. & Novikov, S.S. (1971) Advances in the chemistry of aliphatic *N*-nitrosamines. *Russ. chem. Rev., 40*, 34-50

Gangolli, S.D., Shilling, W.H., Lloyd, A.G. (1974) A method for the destruction of nitrosamines in solution. *Food Cosmet. Toxicol., 12*, 168

Golovanova, O.F., Pepekin, V.I., Korsunskii, B.L., Gafurov, R.G., Eremenko, L.T. & Dubovitskii, F.I. (1974) Kinetic and thermochemical investigation of *N*-nitrosopiperidine. *Bull. Acad. Sci. USSR, Div. chem. Sci.*, 1417-1419

Goodall, C.M. & Kennedy, T.H. (1976) Carcinogenicity of dimethylnitramine in NZR rats and NZO mice. *Cancer Lett., 1*, 295-298

REFERENCES

Gough, T.A., Webb, K.S. & McPhail, M.F. (1978) *Diffusion of nitrosamines through protective gloves.* In: Walker, E.A., Castegnaro, M., Griciute, L. & Lyle, R.E., eds, *Environmental Aspects of N-Nitroso Compounds*, Lyon, International Agency for Research on Cancer (*IARC Scientific Publications* No. 19), pp. 531-534

Gough, T.A., Sugden, K. & Webb, K.S. (1975) Pyridine catalyzed reaction of volatile N-nitrosamines with heptafluorobutyric anhydride. *Analyt. Chem., 47*, 509-512

Gough, T.A., Pringuer, M.A., Sugden, K., Webb, K.S. & Simpson, C.F. (1976) Pyridine catalyzed reaction of N-nitrosodimethylamine with heptafluorobutyric anhydride. *Analyt. Chem., 48*, 583-585

Gough, T.A., Webb, K.S. & Coleman, R.F. (1978) Estimate of the volatile nitrosamine content of U.K. food. *Nature, Lond., 272*, 161-163

Gowenlock, B.G., Pfab, J. & Williams, G.C. (1978) Quantum yields for the photolysis of some nitrosamines in solution. *J. Chem. Res.(S)*, 362-363

Groenen, P.J., de Cock-Bethbeder, M.W., Jonk, R.J.G. & Von Ingen, C. (1976) *Further sutdies on the occurrence of volatile* N-*nitrosamines in meat products by combined gas chromatography and mass spectrometry.* In: Tinbergen, B.J. & Krol, B., eds, *Proc. 2nd Int. Symp. Nitrite Meat Prod., Zeist*, Wageningen, PUDOC, The Netherlands, pp. 227-237

Hanst, P.L., Spence, J.W., Miller, M. (1977) Atmosphere chemistry of N-nitrosodimethylamine. *Environ. Sci. Technol., 11*, 403-405

Hatt, H.H. (1943) *Unsym. dimethylhydrazine hydrochloride.* In: Blatt, A.H., ed., *Organic Synthesis*, Coll. Vol. 2, John Wiley, pp. 211-213

Haszeldine, R.N. & Jander, J. (1954) Studies in spectroscopy, part VI, ultraviolet and infra-red spectra of nitrosamines, nitrites and related compounds. *J. Chem. Soc.*, 691-695

Haszeldine, R.N. & Mattison, B.J.H. (1955) Studies in spectroscopy, part IX, Further studies on nitrosamines and nitrites. *J. Chem. Soc.*, 4172-4185

Heyns, K. & Röper, H. (1971) Specific separation and detection techniques for nitrosamines by a combination of capillary GLC and mass spectrometry. *Z. Lebensm. Unters.-Forsch., 145*, 69-75

Hirano, T. (1973) γ-Radiolysis of dimethylnitrosamine. *Food Irrad., 8*, 70-77

Hünig, S., Geldern, L. & Lücke, E. (1963) O-Alkylnitrosoimmonium salts, a new class of compounds. *Angew. Chem., Int. Ed., 2*, 327-328

Hünig, S., Geldern, L. & Lücke, E. (1963) O-Alkyl-nitrosimmoniumsalze, eine neue Verbindungsklasse. *Angew. Chem., 75*, 476

International Agency for Research on Cancer (1979a) *IARC Monographs on the Evaluation of the Carcinogenic Risk of Chemicals to Humans*, supplement No. 1, Lyon, International Agency for Research on Cancer

International Agency for Research on Cancer (1979b) *Report of the first meeting of the working group on methods for containment, destruction and disposal of carcinogenic waste from laboratories.* IARC Internal Report No. 79/002

Iversen, P.E. (1971) Organic electrosyntheses. III. Reduction of N-nitrosamines. *Acta chem. scand., 25*, 2337-2340

Jakubowski, E. & Wan, J.K.S. (1973) ESR study of UV-irradiated dimethylnitrosamine at 77°K: Evidence of the primary N-N cleavage. *Mol. Photochem., 5*, 439-441

Johnson, E.M. & Walters, C.L. (1971) The specificity of the release of nitrite from nitrosamines by hydrobromic acid. *Analyt. Lett., 4*, 383-386

Jones, E.C.S. & Kenner, J. (1932) The analogy between the benzidine change and the dissociation of oxides of nitrogen. A new reagent for the recovery of secondary bases from nitrosamines and for purifying amines. *J. Chem. Soc.*, 711-715

Jørgensen, K.A., Shabana, R., Scheibye, S. & Lawesson, S.-O. (1980a) *The reaction of N-nitroso compounds with 2,4-bis(4-methoxyphenyl)-1,3,2,4-dithiadiphosphetane-2,4-disulfide. New aspects of the destruction of the N-nitroso function.* In: Walker, E.A., Griciute, L., Castegnaro, M. & Börszönyi, M., eds, N-*Nitroso Compounds: Analysis, Formation and Occurrence*, Lyon, International Agency for Research on Cancer (*IARC Scientific Publications* No. 31), pp. 129-138

Jørgensen, K.A., Shabana, R., Scheibye, S. & Lawesson, S.-O. (1980b) Studies on organophosphorus compounds. XXXII - Reactions of 2,4-bis(4-methoxyphenyl)-1,3,2,4-dithiadiphosphetane-2,4-disulfide with compounds containing the N→O function. *Bull. Soc. chim. Belg., 89*, 247-253

Kano, S., Tanaka, Y., Sugino, E., Shibuya, S. & Hibino, S. (1980) Reductive denitrosation of nitrosamines to secondary amines with metal halide/sodium borohydride. *Synthesis*, 741-742

Katsuhiko, H. & Kinjiro, Y. (1979) Microbial degradation of nitrosamines. I. Inducible breakdown of nitrosamines. *Nippon Suisan Gakkaishi, 45*, 925-928

Kozak, P., Kasparova, Z. & Jurecek, M. (1972) Analytické aspekty oxidace organickych dusikatych Latek Kyselinov chromovov XXV Oxidace N-nitrozo aminů. *Sbornik vedeckych praci, 27*, 3-16

Krüger, F.W., Bertram, B. & Eisenbrand, G. (1976) Metabolism of nitrosamines *in vivo*. V. Investigation on $^{14}CO_2$ exhalation, liver RNA labelling and isolation of two metabolites from urine after administration of (2,50-^{14}C-) dinitrosopiperazine to rats. *Z. Krebsforsch., 85*, 125-134

REFERENCES

Lau, M.P., Cessna, A.J., Chow, Y.L. & Yip, R.W. (1971) Flash photolysis of N-nitrosopiperidine. The reactive transient. *J. Am. Chem. Soc., 93*, 3808-3809

Lehmstedt, K. (1927) Die Bestimmung sekundärer Nitrosamingruppen. *Ber. dt. Chem. Ges., 60*, 1910-1912

Leotte, H. (1964) Nitrosamines, propriétés physico chimiques et utilisation. *Rev. Port. Quim., 6*, 163-166

Lijinsky, W. & Taylor, H.W. (1975) Carcinogenicity of methylated dinitrosopiperazines in rats. *Cancer Res., 85*, 1270-1273

Looney, C.E., Phillips, W.D. & Reilly, E.L. (1957) Nuclear magnetic resonance and infra-red study of hindered rotation in nitrosamines. *J. Am. Chem. Soc., 79*, 6136-6142

Lund, H. (1957) Electroorganic preparations. III. Polarography and reduction of N-nitrosamines. *Acta chem. scand., 11*, 990-996

Lunn, G., Sansone, E.B., Keefer, L.K. (1981) Reductive destruction of N-nitrosodimethylamine as an approach to hazard control in the carcinogenesis laboratory. *Food Cosmet. Toxicol., 19*, 493-494

Mador, I.L. & Rekers, L.J. (1957) Method of preparing acid salts of hydroxylamine. *US Patent No. 2 950 954*

Magee, P.N. & Barnes, J.M. (1967) Carcinogenic nitroso compounds. *Adv. Cancer Res., 10*, 163-246

Magee, P.N., Montesano, R. & Preussmann, R. (1976) N-*Nitroso compounds and related carcinogens*. In: Searle, C.E., ed., *Chemical Carcinogens*, A.C.S. Monogr. Ser. No. 173, American Chemical Society, Washington, DC, pp. 491-625

Meerwein, H. (1966) Triethyloxonium fluoroborate. *Org. Synth., 46*, 113-115

Meerwein, H., Hinz, G., Hofmann, P., Kroning, E. & Pfeil, E. (1937) Über tertiäre Oxoniumsalze. I. *J. prakt. Chem., 147*, 257-285

Meerwein, H., Battenberg, E., Gold, H., Pfeil, E. & Willfang, G. (1939) Über tertiäre Oxoniumsalze. II. *J. prakt. Chem., 154*, 83-156

Michejda, C.J. & Schluenz, R.W. (1973) Reactions of N-nitrosamines with Grignard and lithium reagents. *J. org. Chem., 33*, 2412-2415

Montesano, R.W. & Bartsch, H. (1976) Mutagenic and carcinogenic N-nitroso compounds: possible environmental hazard. *Mutat. Res., 32*, 179, 228

Neurath, G., Pirmann, B. & Dünger, M. (1964) Identification of N-nitroso compounds and asymmetric hydrazines as their 5-nitro-2-hydroxy-benzal derivatives and application on the microscale. *Chem. Ber., 97*, 1631-1638

Overberger, C.G., Lombardino, J.G. & Hiskey, R.G. (1958) Novel reduction of N-nitrosodialkylamines - a new reaction. *J. Am. Chem. Soc., 80*, 3009-3012

Pensabene, J.W., Fiddler, W., Dooley, C.J., Doerr, R.C., Wasserman, A.E. (1972) Spectral and gas chromatographic characteristics of some N-nitrosamines. *J. Agric. Food Chem.*, *20*, 274-277

Polo, J. & Chow, Y.L. (1976) Efficient photolytic degradation of nitrosamines. *J. natl. Cancer Inst.*, *56*, 997-1001

Preussmann, R. & Eisenbrand, G. (1972) Problems and recent results in the analytical determination of N-nitroso compounds. *Topics in Chemical Carcinogenesis*, University of Tokyo Press, pp. 323-339

Pulidori, F., Borghesani, G., Bighi, C. & Pedriali, R. (1970) Reduction mechanism of the nitrogen compound at the DME. I. Di-n-propyl-N-nitrosamine. *J. electroanalyt. Chem.*, *27*, 385-396

Rahman, M.T., Ara, I. & Salahuddin, A.F.M. (1976) Nitroso compounds. 1. The reaction of phenylcopper reagents with nitrosamines (Cleavage of the N-N bond). *Tetrahedron Lett.*, 959-962

Rainey, W.T., Christie, W.H. & Lijinsky, W. (1978) Mass spectrometry of N-nitrosamines. *Biomed. Mass Spectrom.*, *5*, 395-408

Rao, C.N.R. & Bhaskar, K.R. (1969) *Spectroscopy of the nitroso groups*. In: Feuer, H., ed., *The Chemistry of the Nitro and Nitroso Groups*, part 1. *Interscience Publishers*, pp. 144-163

Rao, G.S. (1977) Thin layer chromatographic separation of piperazine and its carcinogenic N-nitroso derivatives. *Sep. Sci.*, *12*, 569-571

Rao, G.S. & MacLennon, D.A. (1977) High pressure liquid chromatographic analysis of carcinogenic N-nitroso derivatives of piperazine resulting from drug-nitrite interactions. *J. analyt. Toxicol.*, *1*, 43-45

Von Rappard, E., Eisenbrand, G. & Preussmann, R. (1976) Selective detection of N-nitrosamines by gas chromatography using a modified microelectrolytic conductivity detector in the pyrolytic mode. *J. Chromat.*, *124*, 247-255

Rounbehler, D.P., Ross, R., Fine, D.H., Iqbal, Z.M. & Epstein, S.S. (1977) Quantitation of dimethylnitrosamine in the whole mouse after biosynthesis *in vivo* from trace levels of precursors. *Science, N.Y.*, *197*, 917-918

Sander, J., Schweinsberg, F., Ladenstein, M., Benzing, H. & Wahl, S.H. (1973) Messung der renalen Nitrosaminausscheidung am Hund zum Nachweis einer Nitrosaminbildung *in vivo*. *Hoppe-Seyler's Z. Physiol. Chem.*, *354*, 384-390

Sander, J., Lahar, J., Ladenstein, M. & Schweinsberg, F. (1974) *Quantitative measurements of* in vivo *nitrosamine formation*. In: Bogovski, P. & Walker, E.A., eds, N-*Nitroso Compounds: Analysis and Formation*, Lyon, International Agency for Research on Cancer (*IARC Scientific Publications* No. 9), pp. 123-131

REFERENCES

Sansone, E.B. & Tewari, Y.B. (1978) *The permeability of laboratory gloves to selected nitrosamines.* In: Walker, E.A., Castegnaro, M., Griciute, L. & Lyle, R.E., eds, *Environmental Aspects of N-Nitroso Compounds*, Lyon, International Agency for Research on Cancer (*IARC Scientific Publications* No. 19), pp. 517-529

Saunders, D.G. & Mosier, J.W. (1980) Photolysis of N-nitrosodi-n-propylamine in water. *J. agric. Food Chem.*, *28*, 315-319

Saunders, D.G., Mosier, J.W., Gray, J.E. & Loh, A. (1979) Distribution movement, and dissipation of N-nitrosodipropylamine in soil. *J. agric. Food Chem.*, *27*, 584-589

Schmidpeter, A. (1963) Reaktionen von Nitrosaminen mit Elektrophilen, I. Die Alkylierung von Nitrosaminen. *Tetrahedron Lett.*, 1421-1424

Schmidpeter, A. (1964) Alkoxydiazenium salts. *Angew. Chem. Int. Ed.*, *3*, 151

Schueler, F.W. & Hanna, C. (1951) A synthesis of unsymmetrical dimethylhydrazine using lithium aluminium hydride. *J. Am. Chem. Soc.*, *73*, 4996

Sen, N.P. (1970) Gas liquid chromatographic determination of dimethylnitrosamine as dimethylnitramine at picogram levels. *J. Chromat.*, *51*, 301-304

Sen, N.P. & Donaldson, B. (1974) *The effect of ascorbic acid and glutathione on the formation of nitrosopiperazines from piperazine adipate and nitrite.* In: Bogovski, P. & Walker, E.A., eds, *N-Nitroso Compounds: Analysis and Formation*, Lyon, International Agency for Research on Cancer (*IARC Scientific Publications* No. 9), pp. 103-106

Smith, P.A.S. (1966) N-*Nitroso compounds*. In: *The Chemistry of Open-Chain Organic Nitrogen Compounds*, Vol. 2, W.A. Benjamin, Inc., N.Y., pp. 470-513

Smith, G.W. & Thatcher, D.N. (1962) Catalytic hydrogenation of nitrosamines to unsymmetrical hydrazines. *Ind. Engng Chem., Product Res. Dev.*, *1*, 117-120

Smyth, W.F., Watkiss, P., Burmicz, J.S. & Hanley, H.O. (1975) A polarographic and spectral study of some C- and N-nitroso compounds. *Analyt. Chim. Acta*, *78*, 81-92

Snider, B.G. & Johnson, D.C. (1979) A photo-electroanalyzer for determination of volatile nitrosamines. *Analyt. Chim. Acta*, *106*, 1-13

Spiegelhalder, B., Eisenbrand, G. & Preussmann, R. (1980) *Occurrence of volatile nitrosamines in food: a survey of the West German market.* In: Walker, E.A., Griciute, L., Castegnaro, M. & Börszönyi, M., eds, N-*Nitroso Compounds: Analysis, Formation and Occurrence*, Lyon, International Agency for Research on Cancer (*IARC Scientific Publications* No. 31), pp. 467-479

Spiegelhalder, B. & Preussmann, R. (1982) *Nitrosamines in rubber*. In: Bartsch, H., O'Neill, I.K., Castegnaro, M. & Okada, M., eds, N-*Nitroso Compounds: Occurrence and Biological Effects*, Lyon, International Agency for Research on Cancer (*IARC Scientific Publications* No. 41) (in press)

Stephany, R.W. & Schuller, P.L. (1980) Daily dietary intakes of nitrate, nitrite and volatile N-nitrosamines in the Netherlands using the duplicate portion sampling technique. *Oncology, 37*, 203-210

Tate, R.L. & Alexander, M. (1975) Stability of nitrosamines in samples of lake water, soil and sewage. *J. natl. Cancer Inst., 54*, 327-330

Tate, R.L. & Alexander, M. (1976) Resistance of nitrosamines to microbial attack. *J. environ. Qual., 5*, 131-133

Telling, G.M. (1972) A gas liquid chromatographic procedure for the detection of volatile N-nitrosamines at the ten parts per billion level in foodstuffs after conversion to their corresponding nitramines. *J. Chromat., 73*, 79-87

Vasundhara, T.S., Jayaranam, S. & Parihar, D.B. (1975) The estimation of N-nitrosamines in tropical regions by reversed-phase paper and thin-layer chromatography. *J. Chromat., 115*, 535-541

Vogel, A.I., Cresswell, W.T., Jeffery, G.H. & Leicester, J. (1952) Physical properties and chemical constitution. Part XXIV. *J. Chem. Soc.*, pp. 514-549

Walker, E.A. (1980) *Some chemical carcinogens and their analysis*. In: Knapman, C.E.H., ed., *Developments in Chromatography-2*, London, Applied Sciences, pp. 69-106

Walker, E.A. & Castegnaro, M. (1974) *A report on the present status of a collaborative study of methods for trace analysis of volatile nitrosamines*. In: Bogovski, P. & Walker, E.A., eds, N-*Nitroso Compounds: Analysis and Formation*, Lyon, International Agency for Research on Cancer (*IARC Scientific Publications* No. 14), pp. 77-83

Walker, E.A., & Castegnaro, M. (1976) *New data on collaborative studies on analysis of volatile nitrosamines*. In: Walker, E.A., Bogovski, P. & Griciute, L., eds, *Environmental* N-*Nitroso Compounds: Analysis and Formation*, Lyon, International Agency for Research on Cancer (*IARC Scientific Publications* No. 14), pp. 77-83

Walker, E.A., Castegnaro, M., Garren, L. & Pignatelli, B. (1978) *Limitations to the protective effect of rubber gloves for handling nitrosamines*. In: Walker, E.A., Castegnaro, M., Griciute, L. & Lyle, R.E., eds, *Environmental Aspects of* N-*Nitroso Compounds*, Lyon, International Agency for Research on Cancer (*IARC Scientific Publications* No. 19), pp. 535-543

REFERENCES

Walters, C.L., Fueggle, D.G. & Lunt, T.G. (1974a) *The determination of total non-volatile nitrosamines in microgram amounts*. In: Bogovski, P. & Walker, E.A., eds, N-*Nitroso Compounds: Analysis and Formation*, Lyon, International Agency for Research on Cancer (*IARC Scientific Publications* No. 9), pp. 22-25

Walters, C.L., Fueggle, D.G. & Lunt, T.G. (1974b) *The determination of total non-volatile nitrosamines in microgram amounts*. In: Kroll, B., ed., *Proc. Int. Symp. Nitrite Meat Prod., Zeist, 1973*, Wageningen, PUDOC, The Netherlands, pp. 53-58

Wasserman, A.E. (1972) *A survey of analytical procedures for nitrosamines*. In: Bogovski, P., Preussmann, R. & Walker, E.A., eds, N-*Nitroso Compounds: Analysis and Formation*, Lyon, International Agency for Research on Cancer (*IARC Scientific Publications* No. 3), pp. 10-15

Watanabe, N., Haruta, M. & Chang, B. (1972) Electrochemical fluorination of N-nitrosodiethylamine. *Bull. Chem. Soc. Japan, 45*, 1275-1281

Waters, W.A. (1978) A new group of nitroxide free radicals formed from aliphatic nitrosamines and thiols. *J. Chem. Soc., Chem. Comm.*, 741-742

Whitnack, G.C., Weaver, R.D. & Krase, H.W. (1963) Polarographic behaviour and large scale electrolysis of some alkylnitrosamines. *NOTS-TP-3253, AD No. 413 029*, pub. U.S. Naval Ordnance Test Station, California

Wolfram, J.H. (1975) A method for the identification and quantitation of volatile nitrosamines. *Diss. Abstr. Int. B, 36B*, 1205

Yoneda, F., Senga, K. & Nishigaki, S. (1973) Dimethylamination and nitrosation of pyrimidines with N-nitrosodimethylamine. *Chem. pharm. Bull., 21*, 260-263

Young, J.C. (1976) Detection and determination of N-nitrosamines by thin-layer chromatography using fluorescamine. *J. Chromat., 124*, 17-28

PUBLICATIONS OF THE INTERNATIONAL AGENCY FOR RESEARCH ON CANCER

SCIENTIFIC PUBLICATIONS SERIES

No. 1 LIVER CANCER (1971)
176 pages   US$ 10.00; Sw. fr. 30.–

No. 2 ONCOGENESIS AND HERPESVIRUSES (1972)
Edited by P.N. Biggs, G. de Thé & L.N. Payne
515 pages   US$ 25.00; Sw. fr. 100.–

No. 3 N-NITROSO COMPOUNDS – ANALYSIS AND FORMATION (1972)
Edited by P. Bogovski, R. Preussmann & E.A. Walker
140 pages   US$ 6.25; Sw. fr. 25.–

No. 4 TRANSPLACENTAL CARCINOGENESIS (1973)
Edited by L. Tomatis & U. Mohr
181 pages   US$ 12.00; Sw. fr. 40.–

No. 5 PATHOLOGY OF TUMOURS IN LABORATORY ANIMALS. VOLUME I TUMOURS OF THE RAT PART 1 (1973)
Editor-in-Chief V.S. Turusov
216 pages   US$ 15.00; Sw. fr. 50.–

No. 6 PATHOLOGY OF TUMOURS IN LABORATORY ANIMALS. VOLUME I TUMOURS OF THE RAT PART 2 (1976)
Editor-in-Chief V.S. Turusov
315 pages   US$ 35.00; Sw. fr. 90.–

No. 7 HOST ENVIRONMENT INTERACTIONS IN THE ETIOLOGY OF CANCER IN MAN (1973)
Edited by R. Doll & I. Vodopija
464 pages   US$ 40.00; Sw. fr. 100.–

No. 8 BIOLOGICAL EFFECTS OF ASBESTOS (1973)
Edited by P. Bogovski, J.C. Gilson, V. Timbrell & J.C. Wagner
346 pages   US$ 32.00; Sw. fr. 80.–

No. 9 N-NITROSO COMPOUNDS IN THE ENVIRONMENT (1974)
Edited by P. Bogovski & E.A. Walker
243 pages   US$ 20.00; Sw. fr. 50.–

No. 10 CHEMICAL CARCINOGENESIS ESSAYS (1974)
Edited by R. Montesano & L. Tomatis
230 pages   US$ 20.00; Sw. fr. 50.–

No. 11 ONCOGENESIS AND HERPESVIRUSES II (1975)
Edited by G. de Thé, M.A. Epstein & H. zur Hausen
Part 1, 511 pages, US$ 38.00; Sw. fr. 100.–
Part 2, 433 pages, US$ 30.00; Sw. fr. 80.–

No. 12 SCREENING TESTS IN CHEMICAL CARCINOGENESIS (1976)
Edited by R. Montesano, H. Bartsch & L. Tomatis
666 pages   US$ 48.00; Sw. fr. 120.–

No. 13 ENVIRONMENTAL POLUTION AND CARCINOGENIC RISKS (1976)
Edited by C. Rosenfeld & W. Davis
454 pages   US$ 20.00; Sw. fr. 50.–

No. 14 ENVIRONMENTAL N-NITROSO COMPOUNDS – ANALYSIS AND FORMATION (1976)
Edited by E.A. Walker, P. Bogovski & L. Griciute
512 pages   US$ 45.00; Sw. fr. 110.–

No. 15 CANCER INCIDENCE IN FIVE CONTINENTS, VOL. III (1976)
Edited by J. Waterhouse, C.S. Muir, P. Correa & J. Powell
584 pages   US$ 40.00; Sw. fr. 100.–

No. 16 AIR POLLUTION AND CANCER IN MAN (1977)
Edited by U. Mohr, D. Schmähl & L. Tomatis
331 pages   US$ 35.00; Sw. fr. 90.–

No. 17 DIRECTORY OF ON-GOING RESEARCH IN CANCER EPIDEMIOLOGY (1977)
Edited by C.S. Muir & G. Wagner
599 pages   US$ 10.00; Sw. fr. 25.–

No. 18 ENVIRONMENTAL CARCINOGENS SELECTED METHODS OF ANALYSIS
Editor-in-Chief H. Egan
Vol. 1-ANALYSIS OF VOLATILE NITROSAMINES IN FOOD (1978)
Edited by R. Preussmann, E.A. Walker, M. Castegnaro & E.A. Wasserman
218 pages   US$ 45.00; Sw. fr. 90.–

No. 19 ENVIRONMENTAL ASPECTS OF N-NITROSO COMPOUNDS (1978)
Edited by E.A. Walker, M. Castegnaro, L. Griciute & R.E. Lyle
566 pages   US$ 50.00; Sw. fr. 100.–

No. 20 NASOPHARYNGEAL CARCINOMA: ETIOLOGY AND CONTROL (1978)
Edited by G. de Thé & Y. Ito
610 pages   US$ 60.00; Sw. fr. 100.–

No. 21 CANCER REGISTRATION AND ITS TECHNIQUES (1978)
by R. MacLennan, C.S. Muir, R. Steinitz & A. Winkler
235 pages   US$ 25.00; Sw. fr. 40.–

No. 22 ENVIRONMENTAL CARCINOGENS – SELECTED METHODS OF ANALYSIS
Editor-in-Chief H. Egan
VOL. 2. – VINYL CHLORIDE (1978)
by D.C.M. Squirrell & W. Thain
142 pages   US$ 45.00; Sw. fr. 75.–

No. 24 ONCOGENESIS AND HERPESVIRUSES III (1978)
Edited by G. de Thé, W. Henle & F. Rapp
1102 pages   Part 1   US$ 30.00; Sw. fr. 50.–
             Part 2   US$ 30.00; Sw. fr. 50.–

No. 25 CARCINOGENIC RISKS – STRATEGIES FOR INTERVENTION (1979)
Edited by W. Davis & C. Rosenfeld
283 pages   US$ 30.00; Sw. fr. 50.–

No. 26 DIRECTORY OF ON-GOING RESEARCH IN CANCER EPIDEMIOLOGY (1978)
Edited by C.S. Muir & G. Wagner
550 pages   Sw. fr. 30.–

No. 27 MOLECULAR AND CELLULAR ASPECTS OF CARCINOGEN SCREENING TESTS (1980)
Edited by R. Montesano, H. Bartsch & L. Tomatis
371 pages   US$ 40.00; Sw. fr. 60.–

No. 28 DIRECTORY OF ON-GOING RESEARCH IN CANCER EPIDEMIOLOGY (1979)
Edited by C.S. Muir & G. Wagner
672 pages   Sw. fr. 30.–

No. 29 ENVIRONMENTAL CARCINOGENS – SELECTED METHODS OF ANALYSIS
Editor-in-Chief H. Egan
VOL. 3. – POLYCYCLIC AROMATIC HYDROCARBONS (1979)
Edited by M. Castegnaro, P. Bogovski, H. Kunte & E.A. Walker
240 pages   US$ 30.00; Sw. fr. 50.–

No. 30 BIOLOGICAL EFFECTS OF MINERAL FIBRES (1980)
Editor-in-Chief J.C. Wagner
Volume 1, 494 pages, US$ 35.00; Sw. fr. 60.–
Volume 2, 510 pages, US$ 35.00; Sw. fr. 60.–

No. 31 N-NITROSO COMPOUNDS: ANALYSIS, FORMATION AND OCCURRENCE (1980)
Edited by M. Börzsönyi
830 pages   US$ 60.00; Sw. fr. 100.–

No. 32 STATISTICAL METHODS IN CANCER RESEARCH (1980)
By N.E. Breslow & N.E. Day
VOL. 1. – THE ANALYSIS OF CASE-CONTROL STUDIES
US$ 30.00; Sw. fr. 50.–

No. 33 HANDLING CHEMICAL CARCINOGENS IN THE LABORATORY – PROBLEMS OF SAFETY (1979)
Edited by R. Montesano, H. Bartsch, E. Boyland, G. Della Porta, L. Fishbein, R.A. Griesemer, A.B. Swan & L. Tomatis
32 pages   US$ 8.00; Sw. fr. 12.–

No. 35 DIRECTORY OF ON-GOING RESEARCH IN CANCER EPIDEMIOLOGY (1980)
Edited by C.S. Muir & G. Wagner
660 pages   Sw. fr. 35.–

No. 37 LABORATORY DECONTAMINATION AND DESTRUCTION OF AFLATOXINS B_1, B_2, G_1, G_2 IN LABORATORY WASTES (1980)
Edited by M. Castegnaro, D.C. Hunt, E.B. Sansone, P.L. Schuller, H.P. Van Egmond & E.A. Walker
59 pages   US$ 10.00; Sw. fr. 18.–

No. 38 DIRECTORY OF ON-GOING RESEARCH IN CANCER EPIDEMIOLOGY 1981 (1981)
Edited by C.S. Muir & G. Wagner
696 pages   Sw. fr. 40.–

No. 40 ENVIRONMENTAL CARCINOGENS – SELECTED METHODS OF ANALYSIS
Editor-in-Chief H. Egan
VOL. IV. – ANALYSIS OF AROMATIC AMINES (1981)
347 pages   US$ 30.00; Sw. fr. 60.–

IARC MONOGRAPHS ON THE EVALUATION OF THE CARCINOGENIC RISK OF CHEMICALS TO HUMANS

Volume 1, 1972 (out of print)
Some inorganic substances, chlorinated hydrocarbons, aromatic amines, N-nitroso compounds, and natural products
184 pages   US$ 4.20; Sw. fr. 12.–

Volume 2, 1973
Some inorganic and organometallic compounds
181 pages   US$ 3.60; Sw. fr. 12.–

Volume 3, 1973
Certain polycyclic aromatic hydrocarbons and heterocyclic compounds
271 pages   US$ 5.40; Sw. fr. 18.–

Volume 4, 1974
Some aromatic amines, hydrazine and related substances, N-nitroso compounds and miscellaneous alkylating agents
286 pages   US$ 7.20; Sw. fr. 18.–

Volume 5, 1974
Some organochlorine pesticides
241 pages   US$ 7.20; Sw. fr. 18.–

Volume 6, 1974
Sex hormones
243 pages   US$ 7.20; Sw. fr. 18.–

Volume 7, 1974
Some anti-thyroid and related substances, nitrofurans and industrial chemicals
326 pages   US$ 12.80; Sw. fr. 32.–

Volume 8, 1975
Some aromatic azo compounds
357 pages   US$ 14.40; Sw. fr. 36.–

Volume 9, 1975
Some aziridines, N-, S- and O-mustards and selenium
268 pages   US$ 10.80; Sw. fr. 27.–

Volume 10, 1976
Some naturally occurring substances
353 pages   US$ 15.00; Sw. fr. 38.–

Volume 11, 1976
Cadmium, nickel, some epoxides, miscellaneous industrial chemicals, and general considerations on volatile anaesthetics
306 pages   US$ 14.00; Sw. fr. 34.–

Volume 12, 1976
Some carbamates, thiocarbamates and carbazides
282 pages   US$ 14.00; Sw. fr. 34.–

Volume 13, 1977
Some miscellaneous pharmaceutical substances
255 pages   US$ 12.00; Sw. fr. 30.–

Volume 14, 1977
Asbestos
106 pages   US$ 6.00; Sw. fr. 14.–

Volume 15, 1977
Some fumigants, the herbicides 2,4-D and 2,4,5-T, chlorinated dibenzodioxins and miscellaneous industrial chemicals
354 pages   US$ 20.00; Sw. fr. 50.–

Volume 16, 1977
Some aromatic amines and related nitro compounds – hair dyes, colouring agents and miscellaneous industrial chemicals
400 pages   US$ 20.00; Sw. fr. 50.–

Volume 17, 1978
Some N-Nitroso compounds
365 pages   US$ 25.00; Sw. fr. 50.–

Volume 18, 1978
Polychlorinated biphenyls and polybrominated biphenyls
140 pages   US$ 13.00; Sw. fr. 20.–

Volume 19, 1979
Some monomers, plastics and synthetic elastomers, and acrolein
513 pages   US$ 35.00; Sw. fr. 60.–

Volume 20, 1979
Some halogenated hydrocarbons
609 pages   US$ 35.00; Sw. fr. 60.–

Supplement No. 1, 1979
Chemicals and Industrial Processes Associated with Cancer in Humans
(IARC Monographs 1–20)
71 pages   US$ 6.00; Sw. fr. 10.–

Volume 21, 1979
Sex hormones (II)
583 pages   US$ 35.00; Sw. fr. 60.–

Volume 22, 1980
Some heavy metals and sweetening agents
208 pages   US$ 15.00; Sw. fr. 25.–

Supplement No. 2, 1980
Long-term and short-term screening assays for carcinogens: a critical appraisal
426 pages   US$ 25.00; Sw. fr. 40.–

Volume 23, 1980
Some metals and metallic compounds
438 pages   US$ 30.00; Sw. fr. 50.–

Volume 24, 1980
Some pharmaceuticals drugs
337 pages   US$ 25.00; Sw. fr. 40.–

Volume 25, 1981
Wood, leather and some associated industries
412 pages   US$ 30.00; Sw. fr. 60.–

Volume 26, 1981
Some antineoplastic and immunosuppressive agents
411 pages   US$ 30.00; Sw. fr. 62.–

Non-serial publications

Alcool et Cancer (in French only)
42 pages   Fr. fr. 35.–; Sw. fr. 14.–

Information Bulletin on the Survey of Chemicals Being Tested for Carcinogenicity
No. 8 (1979)
604 pages   US$ 25.00; Sw. fr. 40.–

Information Bulletin on the Survey of Chemicals Being Tested for Carcinogenicity
No. 9 (1981)
294 pages   US$ 20.00; Sw. fr. 41.–

Cancer Morbidity and Causes of Death Among Danish Brewery Workers (1980)
145 pages   US$ 25.00.–; Sw. fr. 45.–